T0296765

COVID-19

COVID-19

The Essentials of Prevention and Treatment

JIE-MING QU
Chief Physician of Pulmonary Diseases and Critical Care Medicine, Director of the Institute of Respiratory Diseases, Medical school of Shanghai Jiao-Tong University, Shanghai, China

BIN CAO
Vice President of the China-Japan Friendship Hospital, Beijing, China
Deputy Dean of the Respiratory Medicine Research Institute, Chinese Academy of Medical Sciences, Beijing, China

RONG-CHANG CHEN
Chief Physician of Pulmonary Diseases and Critical Care Medicine, Director of the Shenzhen Institute of Respiratory Disease, Guangdong, China
Deputy Director of the National Clinical Research Center of Respiratory Disease, Beijing, China

ELSEVIER

Elsevier
Radarweg 29, PO Box 211, 1000 AE Amsterdam, Netherlands
The Boulevard, Langford Lane, Kidlington, Oxford OX5 1GB, United Kingdom
50 Hampshire Street, 5th Floor, Cambridge, MA 02139, United States

Library of Congress Cataloging-in-Publication Data
A catalog record for this book is available from the Library of Congress

British Library Cataloguing-in-Publication Data
A catalogue record for this book is available from the British Library

ISBN: 978-0-12-824003-8

For information on all Elsevier publications
visit our website at https://www.elsevier.com/books-and-journals

Publisher: Glyn Jones
Editorial Project Manager: Naomi Robertson
Production Project Manager: Maria Bernard
Cover Designer: Matthew Limbert

Typeset by SPi Global, India

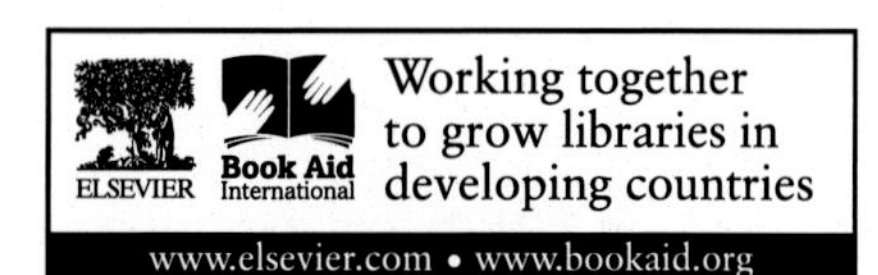

Contents

Author information

Consultants: Nanshan Zhong, Chen Wang
Editors-in-chief: Jieming Qu, Bin Cao, Rongchan Chen
Associate editors-in-chief: Jing Zhang, Yiming Wang, Min Zhou

Editors (listed as the alphabetic order of surname):

Bin Cao (Department of Pulmonary and Critical Care Medicine, China-Japan Friendship Hospital, Beijing, China)

Rong-Chang Chen (Shenzhen Institute of Respiratory Diseases, Shenzhen, Guangdong, China)

Chunhua Chi (Department of General Medicine, Peking University First Hospital)

Qiang Guo (Department of Critical Care Medicine, Suzhou University First Affiliated Hospital, Suzhou, China)

Youmin Guo (Department of imaging, Xi'an Jiaotong University First Affiliated Hospital, Xi'an, China)

Ming Hu (Department of Critical Care Medicine, Wuhan Pulmonary Hospital, Wuhan, China)

Yi Huang (Department of Pulmonary and Critical Care Medicine, Changhai Hospital, Navy Medical University, Shanghai, China)

Rongmeng Jiang (Second Department of Infectious disease, Beijing Ditan Hospital, Beijing, China)

Jian Li (Clinical Research Center, Ruijin Hospital, Shanghai Jiaotong University School of Medicine, Shanghai, China)

Qiang Li (Department of Pulmonary and Critical Care Medicine, Tongji University Shanghai Oriental Hospital, Shanghai, China)

Liang Liu (Tongji Medicolegal Expertise Center in Hubei; Department of Forensic Medicine, Huazhong University of Science and Technology, Wuhan, China)

Jie-Ming Qu (Department of Pulmonary and Critical Care Medicine, Ruijin Hospital, Shanghai Jiaotong University School of Medicine, Shanghai, China)

Yi Shi (Department of Pulmonary and Critical Care Medicine, Jinlin Hospital of Nanjing University, Nanjing, China)

Yuan-Lin Song (Department of Pulmonary and Critical Care Medicine, Zhongshan Hospital, Shanghai Medical College, Fudan University, Shanghai, China)

Anhua Wu (Infection Control Center, Xiangya Hospital of Central South University, Changsha, China)

Chen Wang (Department of Pulmonary and Critical Care Medicine, Center of Respiratory Medicine, China-Japan Friendship Hospital, Beijing, China; National Clinical Research Center for Respiratory Diseases, Beijing, China; Chinese Academy of Medical Sciences and Peking Union Medical College, Beijing, China; Institute of Respiratory Medicine, Chinese Academy of Medical Sciences, Beijing, China)

Yi-Min Wang (Department of Pulmonary and Critical Care Medicine, China-Japan Friendship Hospital, Beijing, China)

Guang Zeng (Chinese Center for Disease Control and Prevention)

Jing Zhang (Department of Pulmonary and Critical Care Medicine, Zhongshan Hospital, Shanghai Medical College, Fudan University, Shanghai, China)

Xiaochun Zhang (Department of Imaging, Zhongnan Hospital of Wuhan University)

Jian-Ping Zhao (Department of Pulmonary and Critical Care Medicine, Tongji Hospital, Tongji Medical College of Huazhong University of Science and Technology, Wuhan, Hubei, China)

Min Zhou (Department of Pulmonary and Critical Care Medicine, Ruijin Hospital, Shanghai Jiaotong University School of Medicine, Shanghai, China)

Prologue

In late 2019 and early 2020, several cases of unexplained pneumonia with fever, cough, fatigue, and/or respiratory distress as the main symptoms appeared in Wuhan, China, within a short time. It was determined that the disease was caused by a new coronavirus—2019 novel coronavirus (2019-nCoV)—and the disease was correspondingly named 2019 coronavirus disease (COVID-19). In the early stage of the epidemic, information on the epidemiological characteristics of COVID-19, clinical treatment options, and protection of medical personnel was rapidly changing and confusing. Therefore, the Chinese Thoracic Society and Chinese Association of Chest Physicians collected and followed up on questions of close concern to frontline medical personnel, and initiated a series of lectures to answer these questions on the "Respiratory Community" platform. The series of 20 lectures brought together experts in the field of respiratory and critical care medicine led by academician Chen Wang, Professor Jie-ming Qu, and experts in other professional fields closely related to the prevention and control of the epidemic, such as Professor Guang Zeng from the Chinese Center for Disease Control and Prevention.

Looking back at the fight against epidemics in China, we can see that with the accumulation of data, the understanding of COVID-19 has become more and more in-depth and comprehensive; accordingly, the prevention and control measures and treatment protocols have been continuously adjusted and improved. In order to reproduce realistically our understanding of COVID-19 and to summarize the prevention and control experience, this book presents the highlights of the lecture series in a question-and-answer format for the benefit of medical staff and readers alike. To date, our understanding of COVID-19 is still relatively superficial, new knowledge and technologies are still being enriched, and the fight against the epidemics has entered a new phase. It is believed that the dynamic perspective on emerging diseases, especially acute respiratory infectious diseases with significant socioeconomic impact, will also shed light on future prevention and control efforts.

On the occasion of publishing this book, we would like to thank all the experts who were invited to participate in the lecture series in the midst of a busy fight against the epidemic, the frontline medical staff for their active participation and interaction, and the organization of the "Respiratory

Community" live platform. Without first-line practical experience, without a summary of research, without discussion, exchange, or even debate, a preliminary understanding of the full picture of a new disease in such a short period of time is not possible.

Given that the book is a collection of lectures, there is a lack of scholarly systematization and terminology, and some of the views are only representative of the situation at the time and may have changed over time. In addition, the book was written in a tight time frame, so we would like to invite, and would appreciate, comments and criticisms from readers.

Jie-ming Qu
President,
Chinese Thoracic Society,
April, 2020

Chen Wang
President,
Chinese Association of Chest Physicians

CHAPTER 1

Respiratory virus and COVID-19

Contents

1. Overview of respiratory viruses

Viral infectious diseases remain a major challenge for human health. Following the emergence of a new coronavirus pneumonia, more than 10,000 species of wild viruses have been mentioned by mass media, but only a few are well recognized. In recent decades, human beings have constantly faced the challenge of bacterial and viral infections. The most common pathogens of new infectious diseases are viruses, the latest being COVID-19. Therefore, we should pay close attention to the severity of respiratory virus infection. There are many common viruses that can cause respiratory infections, including influenza-related viruses, human metapneumovirus, measles virus, rhinovirus, enterovirus, coronavirus, respiratory tract syncytial virus, adenovirus, cytomegalovirus, herpes simplex virus, etc. In particular, there are more than 100 species of coronaviruses.

We usually ignore "coronavirus" due to its weak relationship with human beings. However, we became aware of it after the spread of severe acute respiratory syndrome (SARS) and COVID-19. Bats seem to be one of the most capable hosts of coronavirus. To date, there are seven known kinds of contagious coronaviruses, including SARS-CoV-2. Human coronavirus 229E, human coronavirus nl63, and human coronavirus OC43 are common viruses causing human colds, without more serious pathogenicity. Since they are usually self-limiting with no specific treatment, we do not pay much attention to them. Now we have realized that SARS and COVID-19 have a serious influence on human society. There is thus an urgent need to pay more attention to respiratory tract-related virus infection.

COVID-19
https://doi.org/10.1016/B978-0-12-824003-8.00001-2

From the "A" of avian influenza to the "Z" of Zika, new viruses have sprouted in recent years and show the uncertainty of outbreaks: the occurrence of new viruses will not end with the alphabet, and mankind will have to face new challenges in the future. Continuous screening and preclinical research are always required for infectious disease control. It is necessary to launch a global virus genome plan and to distribute an "active attack" and "all-out attack" strategy on new emerging infectious diseases.

When infected by a virus, patients may present with various clinical symptoms, including colds, pharyngitis, tracheitis, bronchiolitis, pneumonia, etc. Although the spectrum of diseases caused by different kinds of virus is diverse, different viruses may result in the same diseases. For example, influenza viruses can cause adult pneumonia; adenovirus can also cause severe pneumonia. Rhinovirus mainly causes colds as well as pneumonia. Infection of cytomegalovirus in an immunocompromised population is a great challenge for modern interventions (high dose chemotherapy, immunosuppressive therapy, organ transplantation, etc.).

Viruses are common pathogens of community acquired pneumonia (CAP), and their importance has been increasingly emphasized. Viruses also play a role in HAP, AECOPD, bronchiectasis, and other diseases.

2. Etiology of COVID-19

The new coronavirus belongs to the beta genus of coronavirus. It has envelopes, round or oval, but is usually polymorphic. Its diameter is 60–140 nm. The World Health Organization (WHO) named it SARS-CoV-2. SARS coronavirus appeared in 2003, originating in Guangdong, Middle East respiratory syndrome (MERS) coronavirus appeared in 2012, originating in the Middle East, and COVID-19 emerged in 2019, originally found in Wuhan. All three of these coronaviruses are contagious and virulent. Four other coronaviruses are also known to cause human diseases. However, they mainly cause colds—accounting for 10%–15% of cold viruses—and the infection is not severe.

Fig. 1.1 shows that the four genera of coronaviruses, alpha, beta, gamma, and delta, have different genetic structures. Even in the same genus, e.g., beta, coronaviruses of different species are quite different. The new coronavirus is quite different from the coronavirus shown in Fig. 1.1.

Fig. 1.2 presents the different genera of coronavirus: the alpha genus is shown in purple, the beta genus in pink, the gamma genus in green, and the delta genus in blue. Gamma is the smallest genus. As shown in Fig. 1.2,

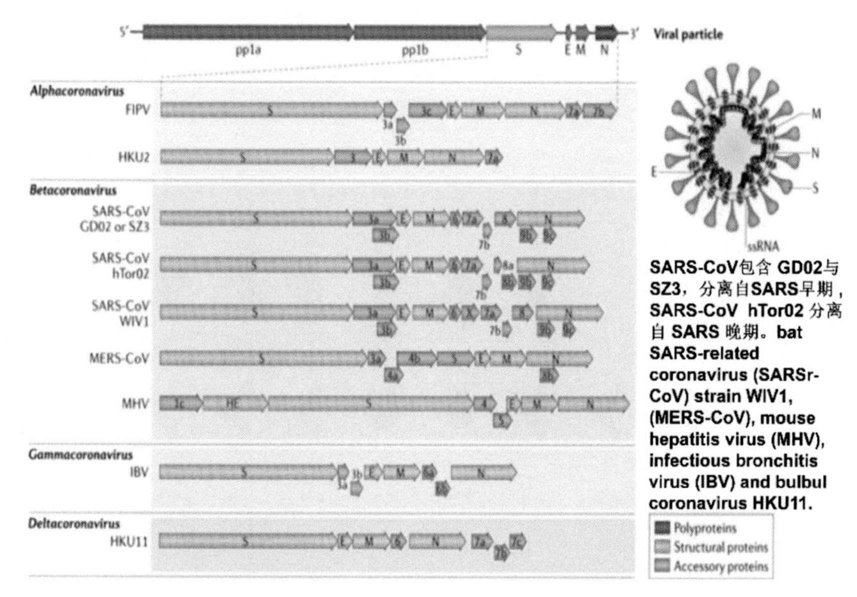

Fig. 1.1 List of coronaviruses: alpha, beta, gamma, and delta.

camels are associated with the Middle East respiratory syndrome virus. Above the camel, the civet cat is shown, which is associated with SARS coronavirus. The bat is the original source of the coronavirus that causes SARS and Middle East respiratory syndrome. The new coronavirus has also been demonstrated to come from bats; specifically, existing evidence suggests that it originates from the chrysanthemum bat. Its intermediate host has not yet been determined, but studies have shown that it may be related to wild animals such as pangolins or snakes.

3. Transmission of COVID-19

First, the source of infection is currently mainly from COVID-19 patients; those with mild and asymptomatic infections can also become a source of infection. The capacity of SARA-CoV-2 infection remains unclear. Based on the pathogenesis of COVID-19, the infectious capacity of SARS-CoV-2 is stronger, especially in severe cases, and the risk of infection is highest during the tracheal intubation intervention process.

Second, there are two main routes of transmission routes for COVID-19: respiratory droplet transmission and close contact transmission.

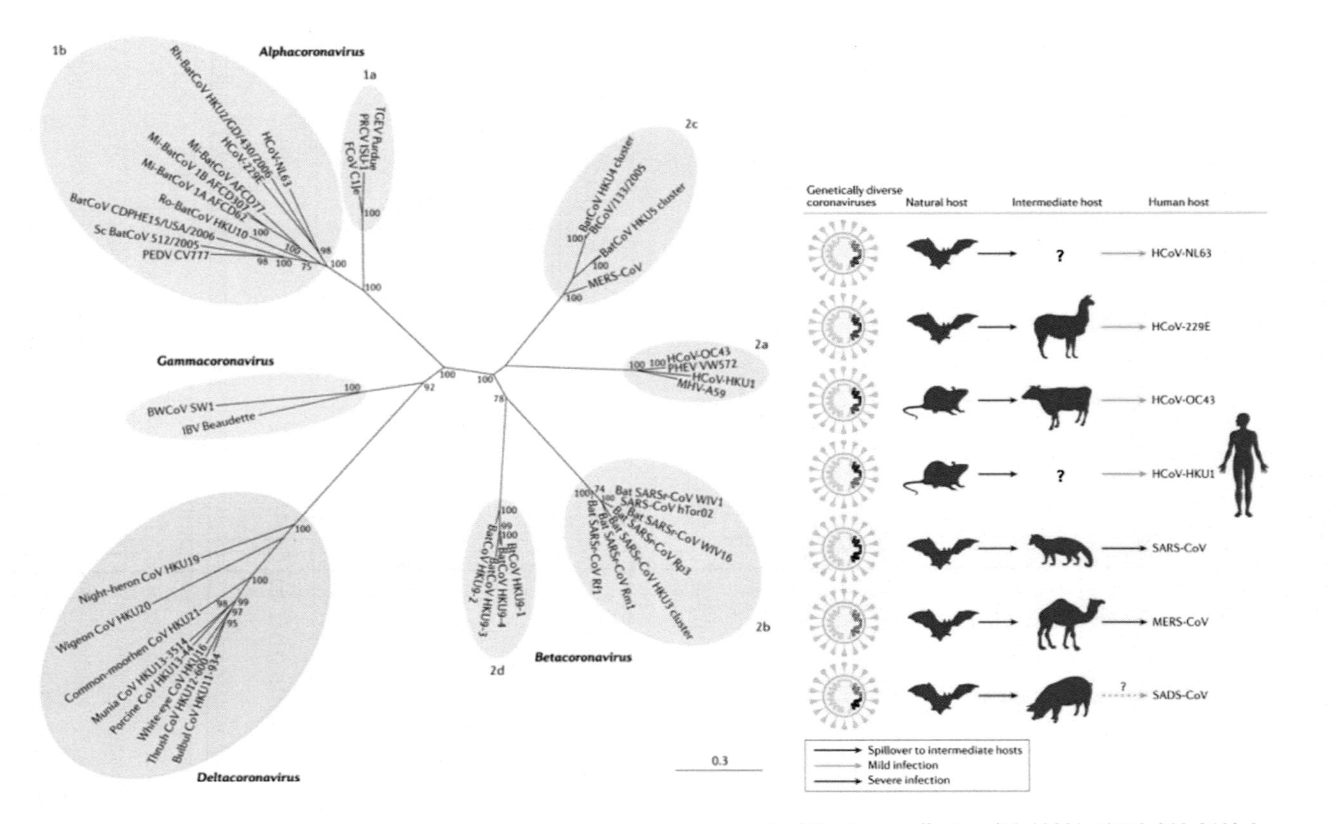

Nat Rev Microbiol 17, 181–192 (2019). https://doi.org/10.1038/s41579-018-0118-9

Fig. 1.2 Coronavirus and intermediate host.

The emergence of respiratory droplets is mainly caused by coughing, sneezing, or talking. The propagation distance for droplets more than 5 μm in diameter is limited and generally less than 1 m. In the case of close contact transmission, droplets can contaminate the surface of objects. The excrement of the patient, such as feces and urine, can pollute the environment as well as the surfaces of objects. If the patient's hands touch the environment or an object's surface of the object, the hands will also be contaminated. The contaminated hands can subsequently make contact with the nasal cavity, oral cavity, or face, which may lead to transmission via close contact. Family clustering transmission is one of the specific transmission characteristics of COVID-19, where there are more than two family members—even up to five members—infected, which confirms the importance of droplet transmission, but does not exclude the possibility of close contact factors.

In a relatively confined space, the virus may be transmitted by aerosol, with a high level of aerosol exposure for a long time. However, there have been very few cases of aerosol transmission until now. New evidence is still needed. Although nucleic acids have occasionally been detected in the air recently, their significance remains to be determined. Generally speaking, the possibility of aerosol propagation is relatively small, and aerosol propagation is not the main route of transmission. It is secondary, but necessary to observe this form of transmission in research. In some patients, the virus can be detected in the feces. It is thus not yet clear whether it can spread through the alimentary canal. At least in some patients we have found that the virus can exist longer in feces than in the respiratory tract, which may bring some challenges for disease prevention and control. More research is needed to determine how long the transmission period is.

The third point to consider regarding transmission is vulnerable people. People are generally susceptible, especially among the immunocompromised population. It is certain that risks are related to exposure mode, quantity, and duration. The elderly and people with underlying diseases are presenting more serious symptoms, while children and infants experience milder effects. No one is innately immune to SARS-CoV-2, even those who have experienced infection. It remains unclear how high the antibody titer level is in the late stage and whether one has the ability to guard against a recurrent infection. In general, the risk of recurrent infection is very low, at least within 6 months to 1 year, with the sustainability of neutralizing antibodies.

The link between the three steps of infectious diseases—source of infection, transmission route, and vulnerable people—is very important.

Infectious diseases can spread only when all three steps exist. The infectious disease will be controlled if any one of these three steps is stopped. How to cut off the transmission channels to prevent the spread of the infection is the most important challenge at this moment. Each infectious disease has its own characteristics. No matter how many doctors, nurses, and beds are available, these are useless if the source of infection or route of transmission is out of control. In the case that no adequate ICU resources are available for the treatment of critically ill patients, people have to face the dilemma of treating or giving up on those patients. COVID-19 is not only a respiratory disease, but also a contagious disease. We hope that we can provide some public education using our knowledge of infectious diseases, so that we can take effective steps to prevent the spread of this disease. That is the only way to control the pandemic. It will definitely be very harsh if we focus only on the patients admitted to hospital.

CHAPTER 2

Pathogenesis of COVID-19

Contents

1. Are there any special concerns on COVID-19 and its pathogenesis?

Combined with clinical manifestations and chest imaging features, such as dry cough and abnormal coagulation function, chest imaging mainly showed multiple small patches and interstitial changes at the early stage, with obvious extravasation and less exudative lesions, which developed into multiple ground-glass opacity and infiltrating shadows in the lungs. In critically ill patients receiving tracheal intubation, infiltration fluid is rare in the trachea, which is different from influenza and avian influenza. We assume that pathogenesis of COVID-19 lung injury could mainly be impairment of the lung interstitium and vascular endothelium. Although ARDS can be found in some patients, exudative lesions are relatively less.

To date, more than 10 studies of COVID-19 patients' body anatomies have been conducted at Wuhan Jin Yin Tan Hospital. Professor Liang Liu from Tongji Medical College of Huazhong University of Science and Technology and academician Xiuwu Bian from the First Affiliated Hospital of the Army Medical University have found that virus particles and inclusion bodies can be seen in the lungs, and common features, such as secondary ARDS, diffuse alveolar injury, and formation of hyaline membrane, are also present. Interestingly, different from other viral pneumonia and bacterial infections, fibrosis changes, with collapsing and occlusion of peripheral small airways as well as macrophages and monocytes infiltration, are common in COVID-19 patients.

COVID-19
https://doi.org/10.1016/B978-0-12-824003-8.00002-4

In viral infection, the pathological change of fibrosis is similar to that of secondary organic pneumonia (SOP). We have called this "quasi-SOP," because further screening and confirmation are required. In clinical practice, similar organic manifestations are commonly seen, either in SARS or in influenza virus infections. When will quasi-SOP occur? According to our existing experience, it often developed in chest imaging at 7–10 days after onset. We are able to hear a "burst sound" by auscultation, which is highly consistent with interstitial lesions. Therefore, glucocorticoid may be helpful. Meanwhile, we should also pay attention to the load of virus as well as the state of humoral immunity. Measures have to be taken after comprehensive consideration. There is no existing evidence from a well-designed RCT study on how to choose the dosage of glucocorticoid. When a widespread epidemic occurs, "ordered" things will be replaced by "disorder" or other disturbing factors, which will affect strong research interpretation. Based on previous experience, intravenous administration of 40–120 mg methylprednisolone is recommended, either as a single dose or multiple doses, and dosage could be adjusted individually. Regarding the usage period, 3 days with a high dose strategy, followed by gradual decrease is suggested by some experience. Currently, facing COVID-19, especially with SOP, the overall treatment strategy remains to be determined with the help of viral load and immune status.

Besides interstitial changes, other features can also be seen within the pulmonary vessels. The description of the vascular endothelium requires further observation by pathologists. However, thrombosis is common, especially hyaline thrombus in the capillaries and pulmonary arterioles. This may be related to the elevation of D-dimer. We are impressed by the thrombus changes within blood vessels.

2. SARS-CoV-2 infection and cardiac injury

In addition to the direct effect on the lungs, the heart is another critical target organ that requires more attention. The virus affects the myocardium as well as the conduction system. This is why some patients experience arrhythmias and other conductive changes. Continuous ECG, myocardial enzymes, and BNP need to be monitored among those patients.

Preliminary studies show that BNP and hypersensitive cardiac troponin are associated with the progress to critical illness. There are still many unknown areas regarding SARS-CoV-2, but evidence has suggested that early deaths occurred in some patients with fulminant myocarditis complicated with sepsis cardiomyopathy, while others suffered septic cardiomyopathy only. The main feature of fulminant myocarditis was cardiomyocyte edema, which required IABP and ECMO to maintain stability. Sepsis cardiomyopathy is a myocardial change of sepsis caused by bacterial infection, with characteristics of left ventricular enlargement and EF downgrading. A recent paper in Intensive Care Med revealed that the proportion of heart failure caused by SARS-CoV-2 infection was as high as 40%. The period from onset of symptoms to death in most patients is 12–24 days, which is not consistent with the development of fulminant myocarditis. The long duration of the disease and delayed treatment will increase the possibility of secondary sepsis cardiomyopathy after mixed infection, which is different from influenza. In some COVID-19 patients, myocardial hypertrophy and chronic ischemic changes are found, which may be a manifestation of coronary heart disease. The longer the duration of the disease, the longer the duration of hypoxemia is maintained, and the more obvious the damage of chronic myocardial ischemia will be. We have observed that most COVID-19 patients died of increased hypersensitive troponin and BNP. In some cases, elevated levels of hypersensitive troponin can be seen in all stages of the disease, ranging from hundreds to thousands ng/ml. However, the relationship with prognosis is still under investigation. Notably, COVID-19 patients can also have acute cardiovascular events, such as acute coronary syndrome (ACS).

3. Other injuries

In addition to the lesions mentioned above, an elevated level of D-dimer can be seen in a large proportion of COVID-19 patients. Several studies have suggested that increased D-dimer is negatively correlated to prognosis. It has a greater possibility to induce systemic inflammatory response. The issues of hypercoagulability and fibrinolysis secondary to hypercoagulability often occur in parallel.

Does DIC occur at the same time? If there is no decrease in fibrinogen consumption and platelets, it should be mainly related to hypercoagulability and fibrinolysis, thrombosis or *in situ* thrombosis formation, or venous thromboembolism (VTE). The fall-off of deep vein thrombosis may lead to blockage of pulmonary veins. It must be confirmed whether the level of D-dimer is high. As learned from anatomical study of the bodies of COVID-19 patients, thrombus can be seen in the middle cerebral artery and pulmonary artery, as well as some hyaline thrombus in peripheral arterioles and capillaries. Therefore, if there is no contraindication, we recommend giving anticoagulation treatment while monitoring the adverse effect of bleeding, especially in elderly patients. Moreover, for patients who have received heparin it is necessary to observe platelet changes and pay attention to the development of HIT.

In addition, acute kidney injury (AKI) has also been found in the anatomy of COVID-19 patients, mainly with renal tubule involvement. It is necessary to monitor renal function in clinical situations.

4. What is the role of cytokine release storm triggered by pathogens in severe and critical cases of COVID-19?

In addition to the presentation of viral infection, cytokine release storm (CRS) induced by pathogens also plays a key role in aggravating clinical symptoms. Recent studies reported that IL-6 and GM-CSF were significantly associated with cytokine release storms. Large amounts of immune cells and tissue fluid are infused within the lungs, demonstrating the existence of a CRS. We have analyzed the temporal curve, comparing the group of surviving patients with deceased patients, and observed the dynamic change of inflammatory mediators and cytokines over time, from no respirator support (such as mask oxygen inhalation) required, to noninvasive ventilation, to invasive mechanical ventilation, or even ECMO. We assume that it would be helpful for us to have a more comprehensive understanding of COVID-19 based on macroscopic and microscopic levels of pathology.

Cytokines involved in CRS induced by different viruses are not identical. H1N1 influenza A virus, H5N1 avian influenza virus, SARS coronavirus, MERS coronavirus, and new SARS-CoV-2 virus are similar, but others are not exactly the same. The proportions of elevated IL-6 and GM-CSF are higher in COVID-19. Further studies on novel biomarkers indicating the development from mild to severe ones, or severe to critical ones, are required. We hope to provide some new targets for the treatment and intervention of COVID-19, which will fundamentally help improve the prognosis as well as reduce the mortality of critical cases.

Removal of cytokines from CRS can be a potential therapeutic target. IL-6, as a major cytokine, plays a critical role in organ damage, and is a potential intervention target. The alternative way is to use an existing IL-6 receptor inhibitor for the treatment of COVID-19. Whether cytokines can be removed by CRRT remains controversial. We believe that with CRRT implementation, the prognosis could be slightly improved. However, whether it is directly related to the mortality rate needs to be studied further.

5. Invading organs and virus-induced infections

The pathophysiological mechanism of COVID-19 is the systemic inflammatory reaction caused by viral pneumonia. We have seen many critically ill patients suffering from shock. However, they actually have only the viral infection instead of bacterial infection. As learned from anatomy studies, we can see the invasion of multiple organs by the virus, presenting with particles or inclusion bodies.

The characteristics of multiple organs involvement of SARS-CoV-2 include viral sepsis. No fever or normal temperature would be one of the special manifestations of viral sepsis. Many severe and critically ill patients in Wuhan showed no fever, which is consistent with viral sepsis. As is commonly known, a variety of viruses can lead to viral sepsis. Forty-two percent of viral infections are mixed infection, such as respiratory syncytial virus, influenza, parainfluenza virus, rhinovirus, and adenovirus. Recent reports also suggested that SARS-CoV-2 was detected along with influenza virus. On the other hand, a meta-analysis did not show the association between of coinfection with aggressive clinical severity. A recent sampling study showed that the proportion of HCMV coinfection is about one-third, that of influenza virus is rare, with no other respiratory syncytial viruses, mycoplasma, and legionella detectable. According to the experience from ICU experts, there seems to be a certain possibility of herpes simplex virus coinfection, but conclusive studies are still in progress.

CHAPTER 3

Clinical features of COVID-19

Contents

1. Clinical manifestation

The incubation period from exposure to symptoms is generally 7–14 days; the shortest is 1 day, the longest is up to 20 days. Fever, fatigue, and dry cough appear to be the most common symptoms at illness onset, but these symptoms, which also present in influenza and other respiratory infections, are nonspecific. Upper respiratory tract symptoms like nasal obstruction and rhinorrhea are relatively rare. In general, the majority of patients have a satisfactory prognosis with a few patients being critically ill. Fatal cases are commonly seen in the elderly and those with chronic underlying diseases, such as diabetes and heart disease.

COVID-19 has several clinical characteristics. (1) It mainly attacks the lungs, but it also frequently involves other organs and systems. (2) Some patients show mild onset symptoms without fever and recover 1 week later. (3) About half of the patients developed dyspnea 1 week after onset, and the symptoms and imaging findings of one-third of sufferers were asynchronous. Some patients initially showed no fever or no obvious coughing or dyspnea, but their chest imaging manifestations continued to progress, and their condition strikingly exacerbated within a week. In severe cases, some quickly deteriorated into acute respiratory distress syndrome (ARDS), septic shock, and metabolic acidosis hard to correct, coagulopathy, and multiple organ failure. (4) Severe or critical patients might present low to moderate fever or even no obvious fever in the course of their illness, which impedes the prevention and control of this epidemic. Therefore, it is pivotal for all hospital departments to screen patients with fever and pneumonia in order to prevent nosocomial infections. (5) Among the patients with symptoms,

COVID-19
https://doi.org/10.1016/B978-0-12-824003-8.00003-6

the incidence rate of severe cases is approximately 20%. As for confirmed cases of COVID-19, it is necessary to analyze and evaluate the severity of the disease, observe closely, and instruct the patients themselves to report any progression in a timely manner. Respiratory failure, a very important indicator of exacerbation, is characterized by symptoms like chest tightness and shortness of breath, and by objective indicators, such as blood oxygen saturation. Time of onset, fever, and chest tightness and shortness of breath are important factors in evaluating the condition. The condition of patients with an onset time of 7–10 days or more and symptoms of chest tightness, shortness of breath, and high fever, is more likely to become aggravated. Persistent high fever and chest tightness are likely to indicate the stage of the peak period. Intervention prior to the peak period in severe patients to prevent the development of severe respiratory failure is vital in treating critically ill patients, because most of these patients will gradually recover if they survive this 2–3-week peak period.

2. Abnormal findings of laboratory tests

The abnormality of COVID-19 in laboratory examination is reflected in several aspects. In the early stages, the white blood cell count is normal or decreased, and the lymphocyte count, an indicator negatively correlated with severity of illness, is reduced. Severe lymphoid depletion ($0.3–0.4 \times 10^9$/L) could be observed in critically ill patients. If the patient's lymphocyte count can gradually recover to more than 1.0×10^9/L, the overall condition will improve accordingly; however, if it fails to rise to 1.0×10^9/L, the state of illness will usually be at a deadlock: although the patient's vital signs can remain stable, it is difficult to achieve the condition of extubation. In addition, the C-reactive protein and erythrocyte sedimentation rate are increased, and procalcitonin (PCT) is normal in most patients. It is necessary to be alert to whether the patient has a bacterial infection in cases with elevated PCT. Often accompanied by circulatory dysfunction or poor distal limb perfusion, a number of severe patients have coagulation dysfunction: in some ICU patients, D-dimers increased substantially, even up to 50 mg/L. Some patients have increased liver enzymes, troponin, and myoglobin, with myoglobin levels above 10,000 μg/L. In addition to D-dimer, liver, and kidney function, we must pay attention to the changes of peripheral blood lymphocytes, especially CD4 + T lymphocytes and NK cells in lymphocyte subsets, the diminishment of which figure is a significant indicator of poor prognosis in sufferers of SARS. If the lymphocyte count of patients with COVID-19 is extremely low—as low as 400/μL or even 200/μL—the chance of a patient's survival will plunge. Given that the attack of the virus on the immune system and the intensity of the body's own excessive immune storms vary, there is a sharp difference in clinical performance of different patients even if there is no change in the virus genotype.

3. Abnormalities and features in imaging

Since the majority of patients present as experiencing pneumonia, imaging examination is very important for COVID-19 diagnosis. In brief, chest imaging in the early stage of infection mainly manifests as multiple small patch lesions and interstitial changes, especially in the lung periphery. As the disease progresses, bilateral lungs show ground-glass opacities (GGOs) and infiltration shadows. Consolidation lesions are common in severe patients, but pleural effusion is rare. Most critically ill patients admitted to ICU develop consolidation in gravity-dependent regions, namely consolidation in lower lobes, which is not readily reversible. Therefore, when pulmonary consolidation is detected in imaging, the disease has already deteriorated and the lungs have been damaged by the virus for quite a long time.

Compared with chest X-ray, more details can be observed in CT images. Therefore, chest CT is highly recommended if condition allows. COVID-19 patients presented with various lesions on chest CT, including GGOs (Fig. 3.1), consolidation (Fig. 3.2), nodules (Fig. 3.3), halo and reversed halo sign (Fig. 3.4), airway changes (Fig. 3.5), etc. Multiple lesions could coexist, presenting a mixed pattern.

Chest CT images manifest different features in COVID-19 patients during different courses of the disease. (1) Early stage: ① Unilateral or bilateral, single or multiple, and focal lesions with a peripheral lung and subpleural distribution are commonly encountered. In the early stage of COVID-19, there are single or multiple and focal GGOs, which are more likely to be unilateral rather than bilateral. ② The lesions could also be massive or small patch GGOs, which are caused by vascular enlargement and the increased number of microvascular, consolidation, nodules, and mosaic sign, which is due to the coexistence of GGOs and air trapping. ③ The lesions sometimes manifest as particular slight GGOs and nodules, and it is easy to miss them in the diagnosis. (2) Progressive stage: ① CT images could present bilateral and multiple GGOs or consolidation with air bronchogram sign, nodules with halo sign, reticular patterns (small vascular networks) inside

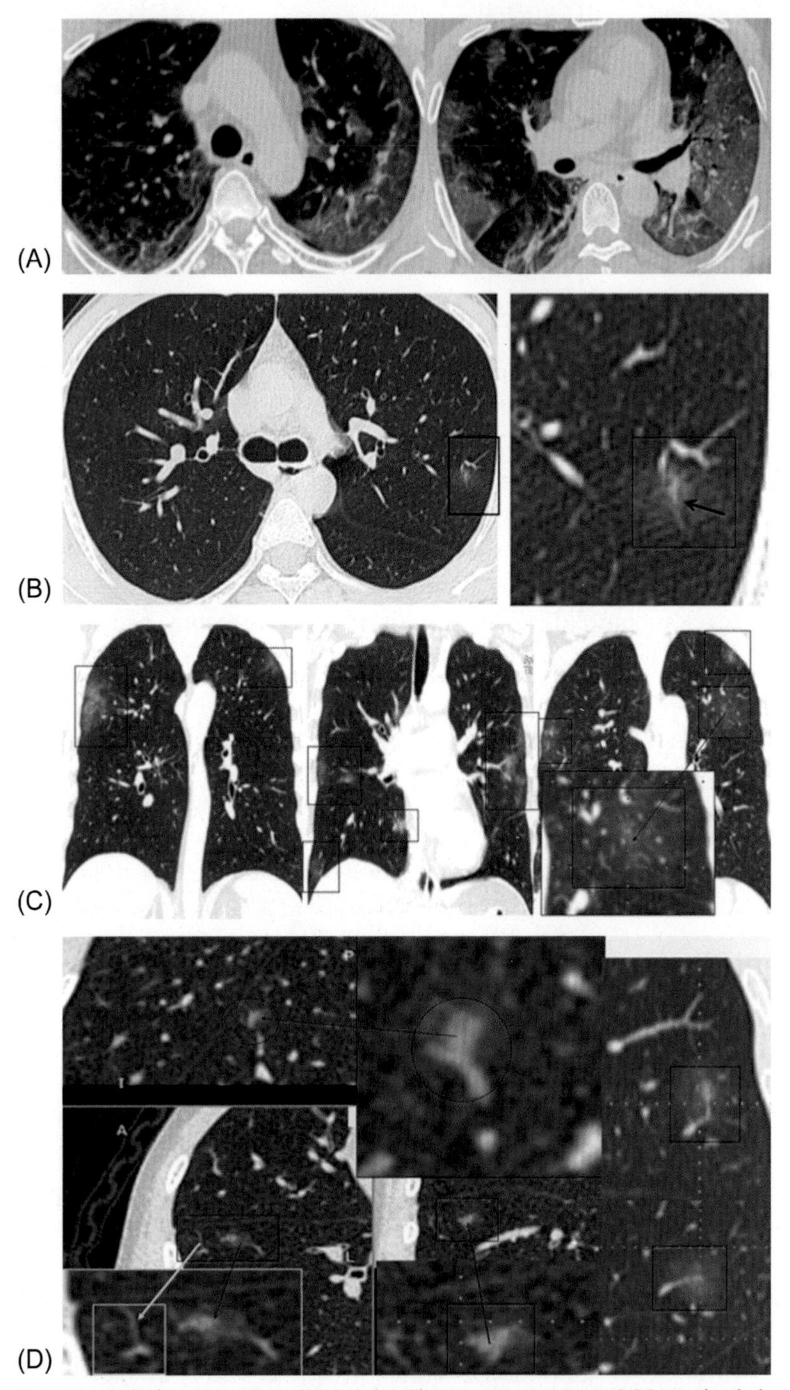

Fig. 3.1 Ground-glass opacities (GGOs) (A) There were massive GGOs in both lungs accompanying reticular patterns. (B) A single GGO with vascular enlargement was found in the left upper lobe and the margin was obscure. (C) There were particular slight GGOs. (D) CT scans showed small vessels enlargement surrounded by GGOs.

(Continued)

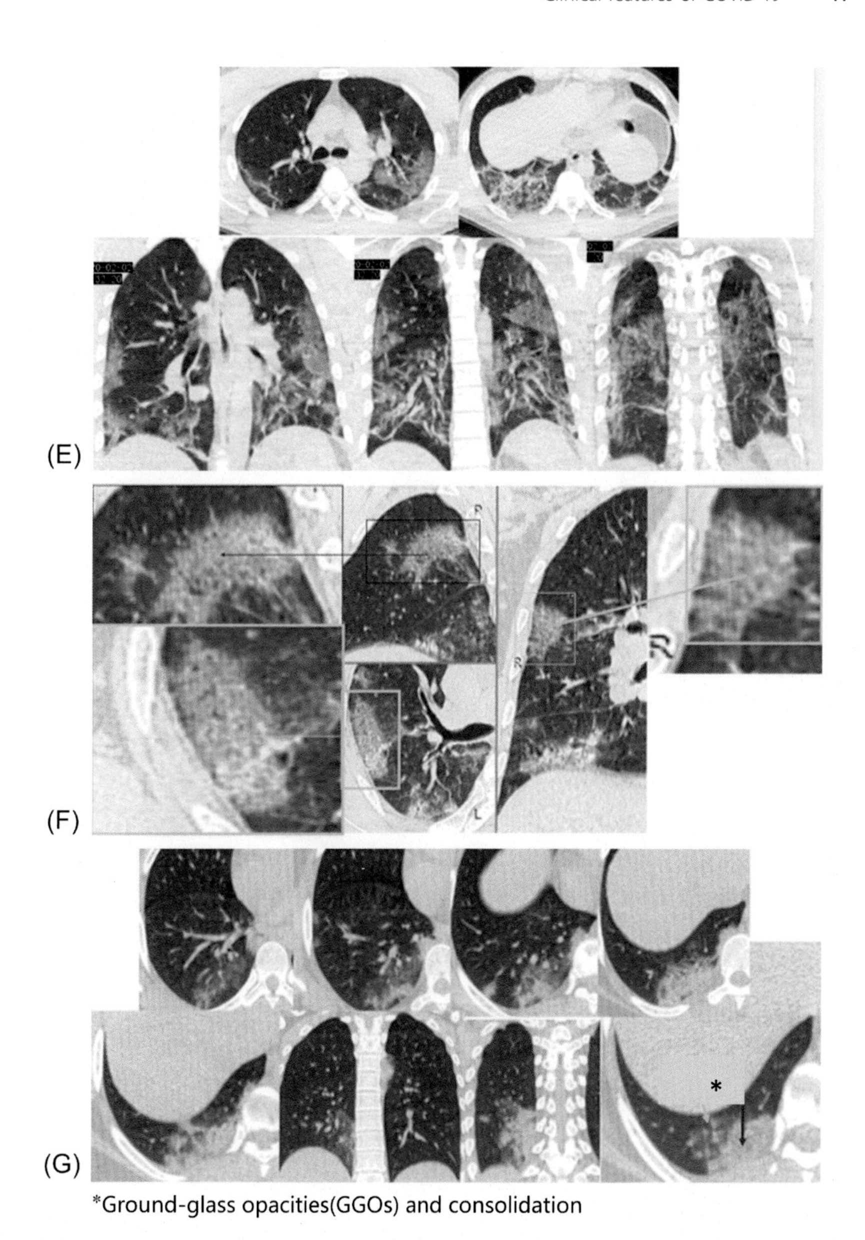

Fig. 3.1, cont'd (E) Multiple GGOs with reticular patterns were observed in both lungs. (F) There were GGOs with "crazy-paving" patterns. (G) There were GGOs with consolidation.

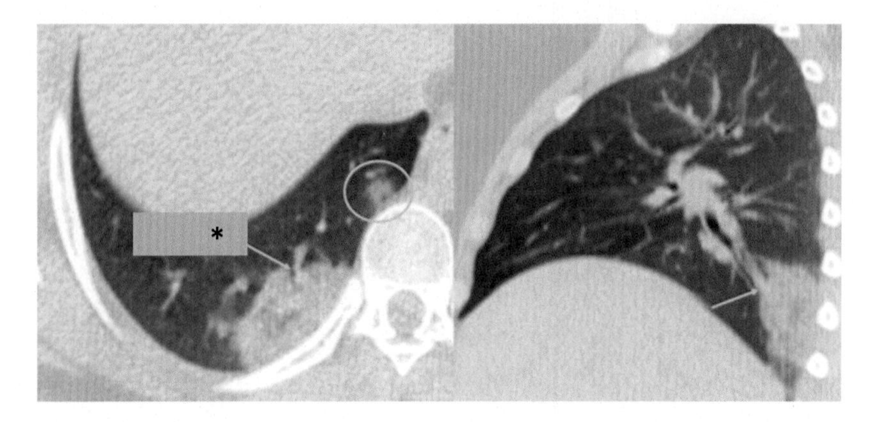

*Consolidation

Fig. 3.2 Consolidation.

lesions, and reversed halo sign. ② New lesions are mainly located in the middle and lower lobes with subpleural distribution, and most manifest as slight GGOs. ③ Other lesions include subsegmental atelectasis and fibrosis. (3) Severe stage: "White lung" with acute lung injury, the range of lesions increasing by 50% in 48 h, pulmonary fibrosis could be seen in CT images of critically ill patients.

The following content introduces radiological features of COVID-19 via etiologically confirmed cases. (1) Imaging findings in the early stage of COVID-19 (Figs. 3.6–3.13). (2) Imaging findings in the progressive stage

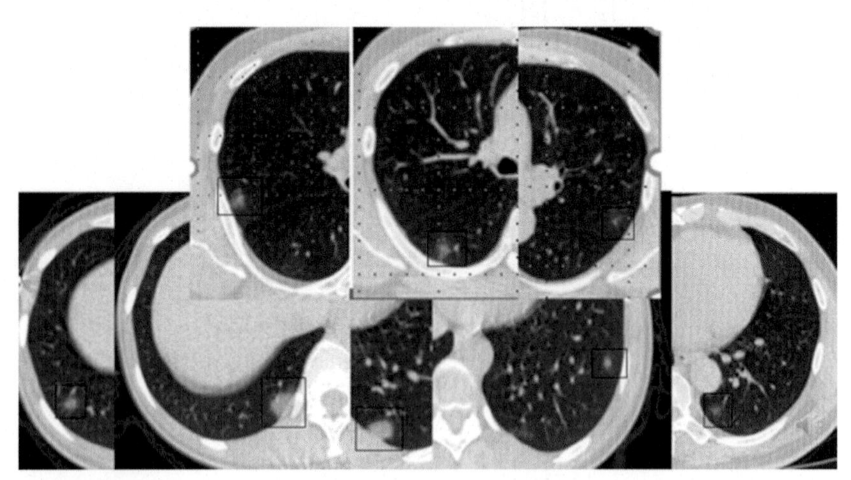

Fig. 3.3 Nodules.

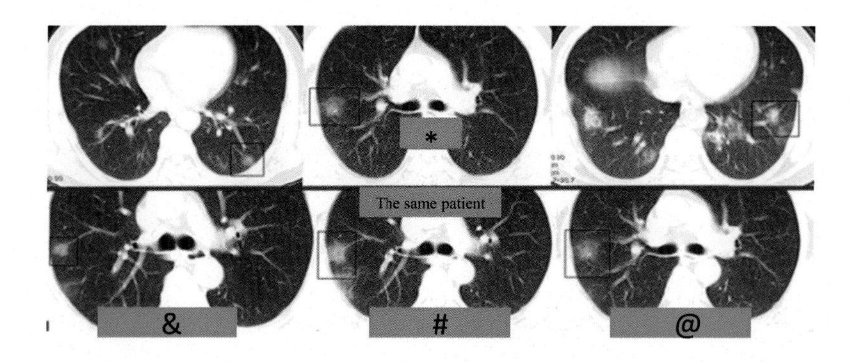

*Halo sign

&Day 1 after beginning with fever

Day 3 after beginning with fever

(A) @Day 8 after beginning with fever

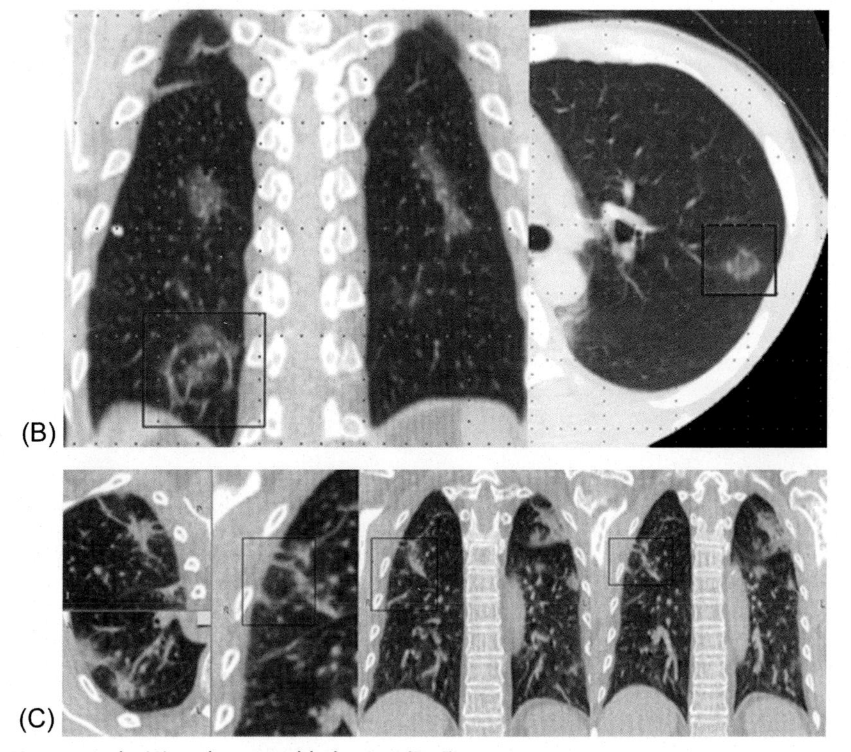

Fig. 3.4 Halo (A) and reversed halo sign (B, C).

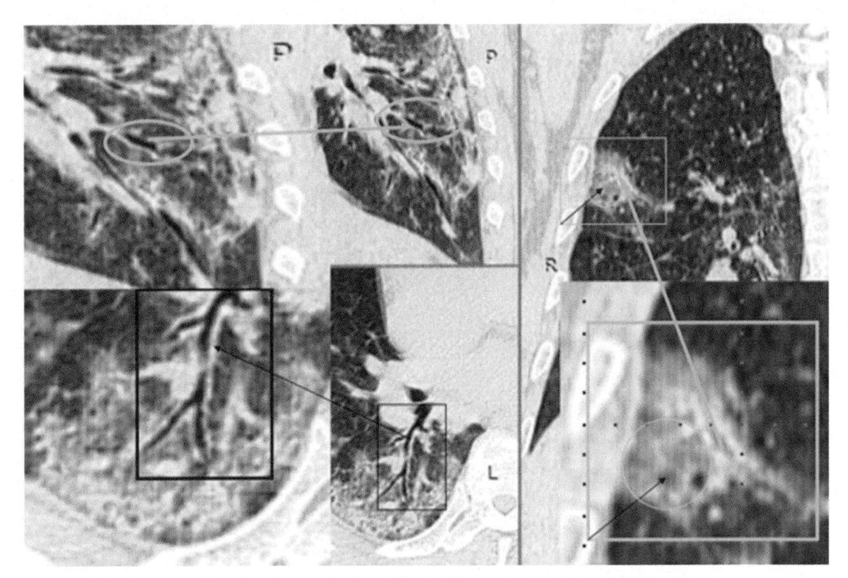

Fig. 3.5 Thickening of the bronchial wall and localized bronchiectasis.

of COVID-19 (Figs. 3.14–3.18). (3) Imported cases of COVID-19: some cases showed evident clustering. The clinical characteristics and CT images of three groups of clustered cases are described as follows. The first group (Figs. 3.19–3.22) contains four confirmed cases with slight imaging changes, which were difficult for chest X-ray to detect. Therefore, for clustered cases,

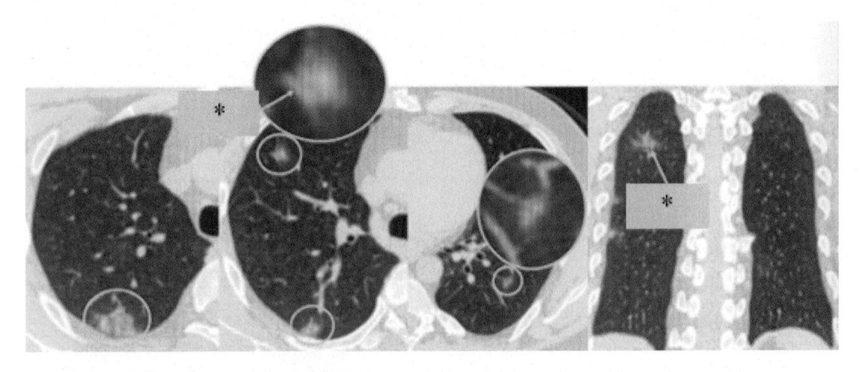

*Lesions with vascular enlargement

Fig. 3.6 Male, 29 years old, with a history of fever for 3 days along with sore throat, but no cough; temperature: 38°C; with epidemic history. Laboratory tests: WBC: 4.62×10^9/L, NEUT% 77.7% , LYMPH% 17.1%. CT scans showed scattered exudative patch lesions, GGOs, and consolidation accompanied by vascular enlargement.

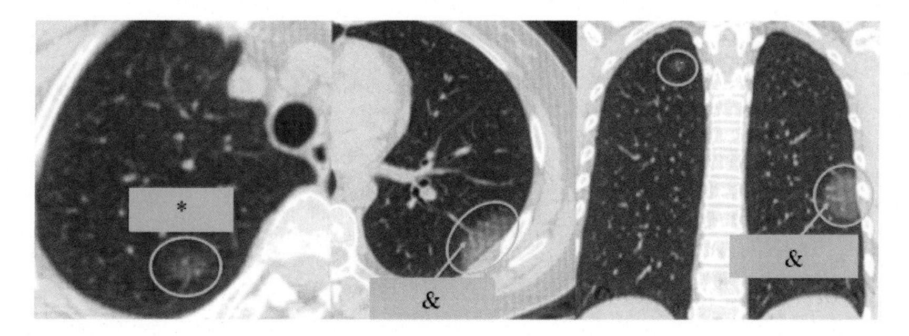

*GGOs

&Lesion with vascular enlargement

Fig. 3.7 Female, 41 years old, with a history of fever for 3 days, but no cough or expectoration; temperature: 38.8°C; with a history of traveling in Wuhan but no contact with anyone with symptoms of COVID-19. Laboratory tests: 3.74×10^9/L， NEUT% 59.7%， LYMPH% 26.5%. CT scans showed there were GGOs in the right upper lobe and left lower lobe, and localized consolidation in the left lower lobe with subpleural distribution.

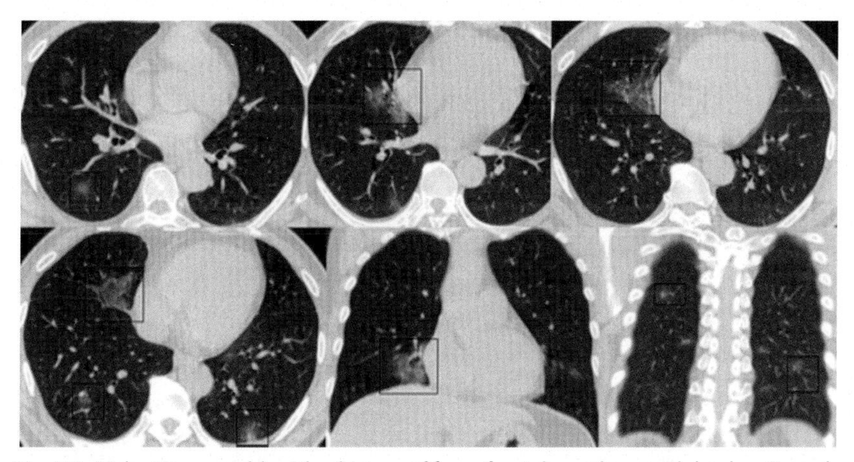

Fig. 3.8 Male, 55 years old, with a history of fever for 4 days, along with back pain, scalp haphalgesia, occasional cough and white sputum, but no sore throat; temperature: 38° C; with a history of contact with confirmed patients. The personal history is hypertension and diabetes. Laboratory tests: WBC 5.73×10^9/L， LYMPH% 12%. CT scans showed multiple patch OGGs located in the subpleural area and interlobar fissure in both lungs, and there were also OGGs in the medial segment of the right middle lobe.

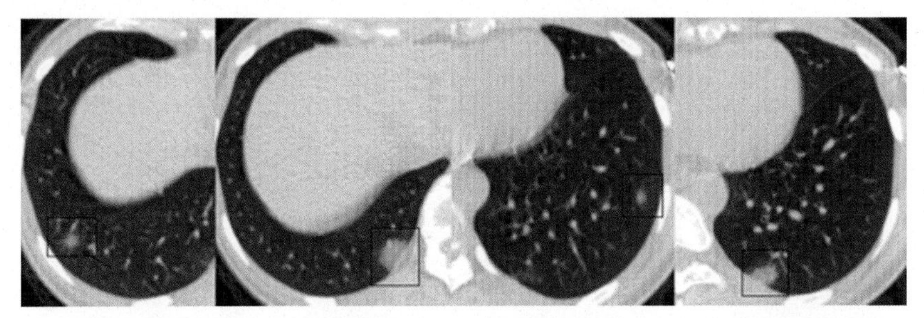

Fig. 3.9 Female, 32 years old, without any obvious symptoms. She was residing in Wuhan and went on an errand on January 21. Her husband had symptoms of COVID-19. Laboratory tests: WBC 4.22×10^9/L, LYMPH% 32.9%. CT scans showed multiple subsolid nodules with intrapulmonary and subpleural distribution in both lungs, surrounded by halo sign.

especially those with epidemiological history, it is necessary to strengthen identification and screening processes.

The second group (Figs. 3.23 and 3.24) contains two confirmed cases, one of which had clinical symptoms but no abnormal changes in imaging examination initially. But as time passed, lesions gradually appeared in chest imaging. Therefore, pathogen detection and follow-up CT scans are necessary for familial clusters even if CT images find nothing. The third group (Figs. 3.25–3.27) contains three confirmed cases. The parents were diagnosed first, while CT images of the son showed suspicious signs of COVID-19 infection, and during disease deterioration, the area of lesions expanded. His first two nucleic acid tests showed negative results, and the third one showed a positive result. Hence, for patients with epidemiological history, especially those with family clustering history, follow-up and repeated performance of nucleic acid tests are needed, even if nucleic acid tests show negative results. (4) For some patients, lesions could expand rapidly in 3–5 days, and even develop fibrosis on the 7th day from the onset of symptoms (Figs. 3.28 and 3.29).

It is notable that we should not concentrate only on pulmonary infiltration shadows, GGOs, features of pneumonia, features of ARDS, etc. Some slight changes in CT images, such as "SOP–like" changes, suggest causes leading to disease progression. As the disease progresses further, we ought to focus not only on infiltration shadows, but also interstitial changes, "SOP–like" changes, changes of pulmonary embolism, and bacterial or fungal infections (Fig. 3.30).

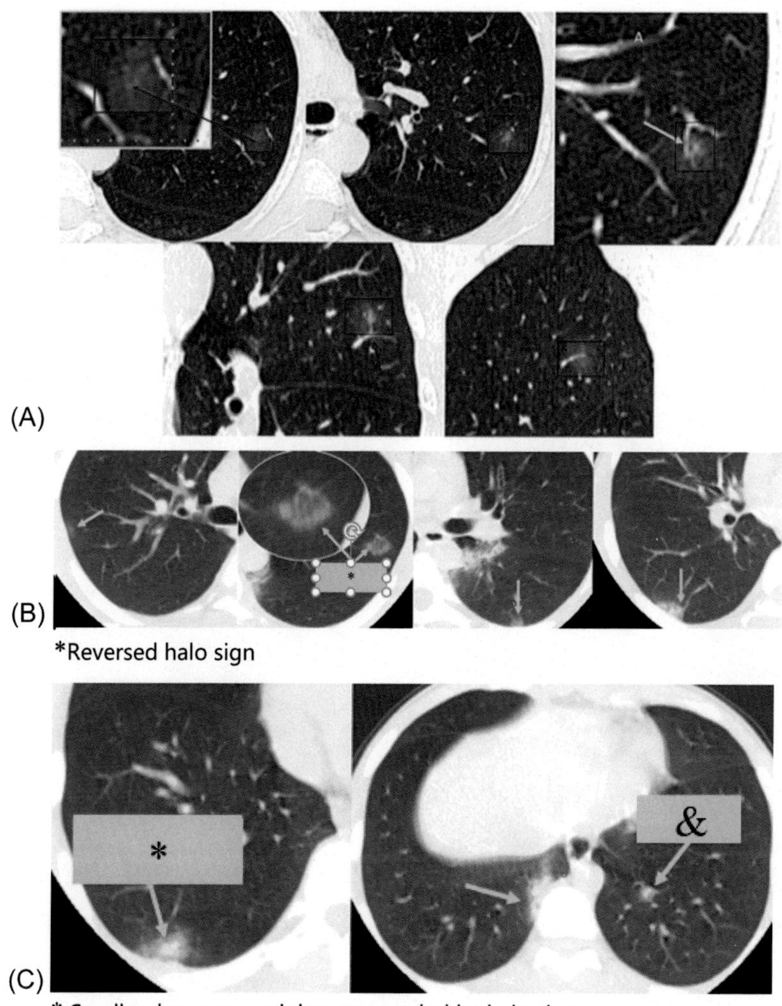

Fig. 3.10 A 22-year-old patient with a history of fever for 1 day, cough and expectoration for 8 days. Temperature: 38.7°C. The patient went to Wuhan on January 10, and left Wuhan on January 21. Laboratory examination: CRP 54.20 mg/L, hs-CRP $>$ 5.00 mg/L, ESR 19 mm/h, D-dimer 2.05 mg/L. CT scans showed small patch GGOs with vascular enlargement. Vascular enlargement could be seen in the slight GGOs (A). An increased number of and enlarged lesions with reversed halo sign (B) and halo sign (C) were observed in follow-up CT scans on January 29, 2020. Small pulmonary nodules were found, most of which were located in the subpleural area.

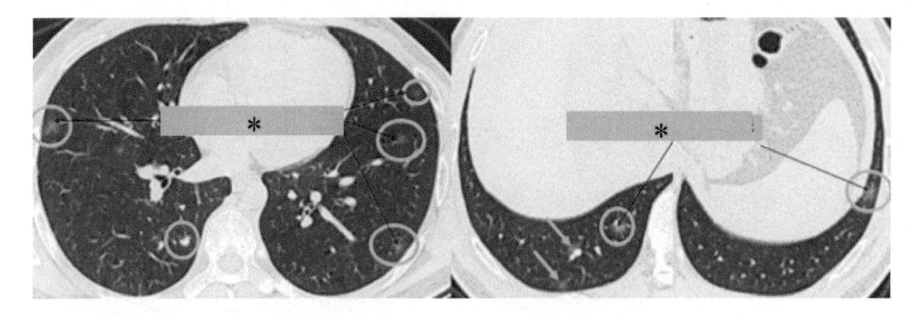

*Small patch GGOs

Fig. 3.11 Male, 26 years old, with a history of fever for 1 day, along with dry cough and fatigue; temperature: 37.1°C. He had returned from Wuhan 3 days previously. Laboratory tests: CRP 9.83 mg/L, hs-CRP > 3.00 mg/L, NEUT% 63.2%, LYM 0.75×10^9/L, ESR 29 mm/h, D-dimer + FDP (−). CT scans showed scattered, multiple, and small patch GGOs and micronodular consolidation.

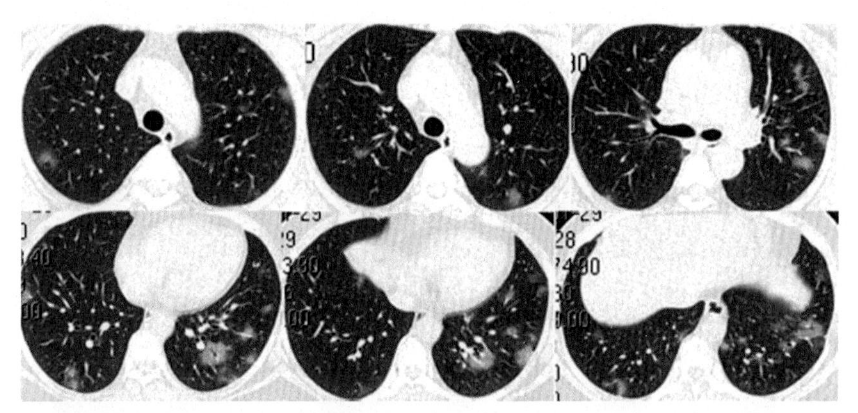

Fig. 3.12 Female, 44 years old, with a history of hyperpyrexia for 4 days, along with cough and expectoration. She returned to her hometown from Hubei Province on January 22, 2020. Laboratory tests: WBC 4.5×10^9/L， NEUT% 57.4%， LYMPH% 36.1%. CT scans showed small patch GGOs, some of which were accompanied by air bronchogram signs and vascular enlargement.

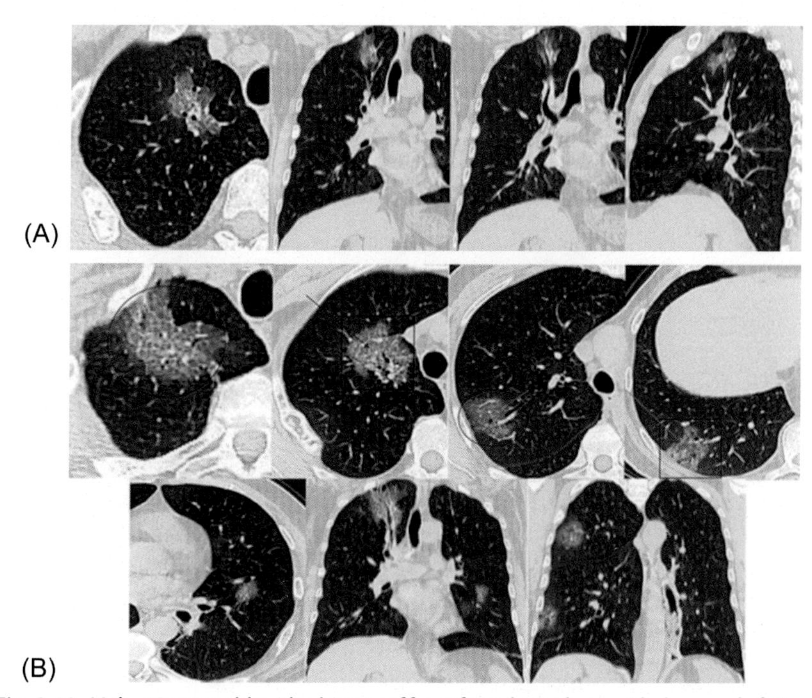

Fig. 3.13 Male, 49 years old, with a history of fever for 4 days, along with dry cough, fatigue, and chest distress; temperature: 38.2°C. He had worked in Wuhan for a long time. He left Wuhan by train on January 18. Laboratory examination: WBC 5.80×10^9/L, NEUT% 76.6%, LYMPH% 15.0%, R 22 times per minute, SpO2 98%. (A) CT scans on January 23 showed GGOs with air bronchogram signs and vascular enlargement in the right upper lobe. (B) Follow-up CT scans on January 26 showed an increased number of and enlarged lesions, and emergence of new ones with "crazy-paving" patterns in the left lung.

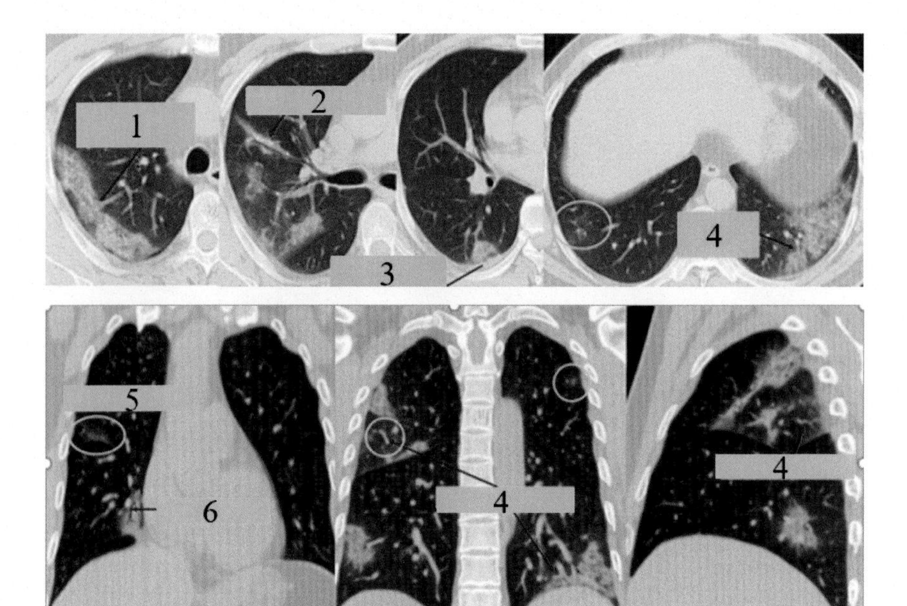

1 GGOs with vascular enlargement
2 Bronchovascular bundles thickening
3 a small amount of pleural effusion
4 Lesions with vascular enlargement
5 GGOs
6 Air bronchogramsign

Fig. 3.14 Male, 47 years old, with a history of fever for 11 days, along with cough and a small amount of white sputum; temperature: 39°C. No travel history to or work in Wuhan, nor contact with anyone who ever stayed in Wuhan. Laboratory tests: WBC 2.92×10^9/L， NEUT% 73.0%， LYMPH% 21.8%. CT scans showed multiple, massive, and patch GGOs and consolidation with a small amount of pleural effusion in the right thorax.

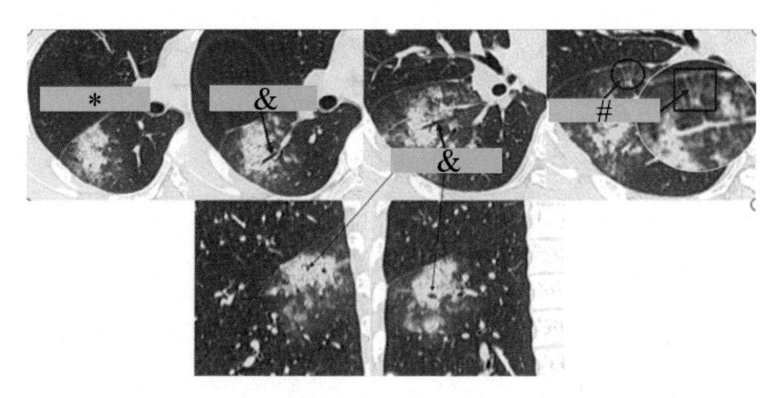

* Lesions surrounded by GGOs
&Air bronchogram sign
(A) # Interlobular septal thickening

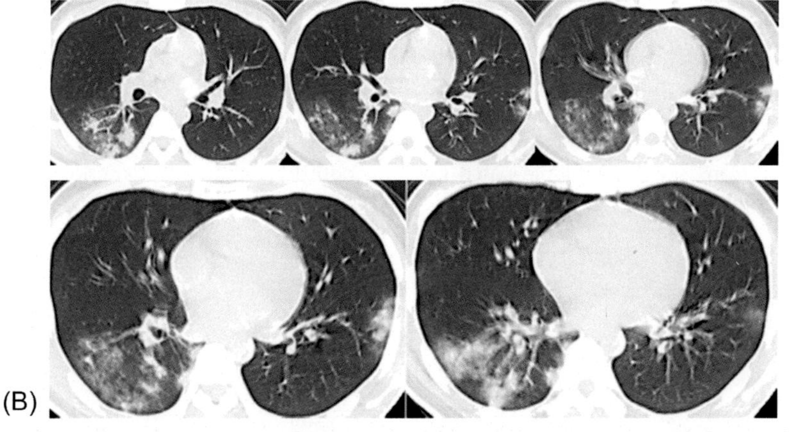

Fig. 3.15 Male, 18 years old, with a history of fever for 1 day, along with fatigue for 6 days and chest distress for 2 days; temperature: 38.5°C. The patient had stayed in Wuhan for 10 days and left on January 22. Laboratory tests: CRP 12.49 mg/L, hs-CRP > 5.00 mg/L, ESR 10 mm/h, D-dimer 1.99 mg/L. (A) CT scans on January 25, 2020 showed massive and mix-pattern consolidation in the right lower lobe, associated with vascular enlargement, air bronchogram, and interlobular septal thickening. In addition, there were GGOs around lesions. (B) Follow-up CT scans on January 29, 2020 showed coexistence of absorption of primary lesions and emergence of enlarged and new ones.

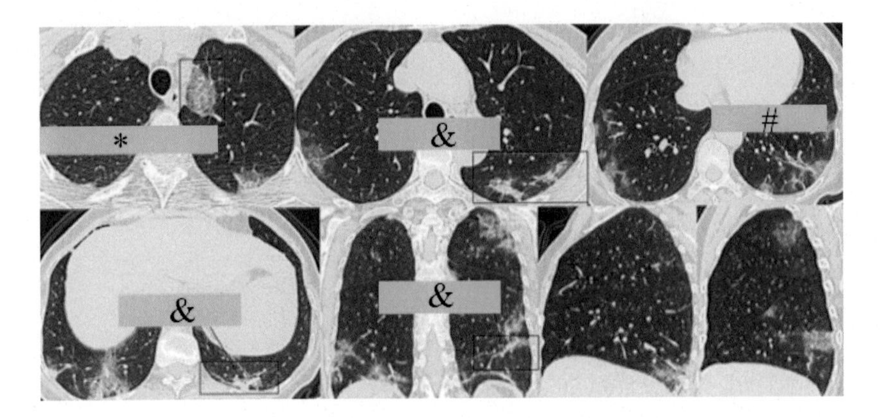

* Lesions with reticular patterns

&Fibrosis lesions

\# Air bronchogram sign

Fig. 3.16 Female, 64 years old, with a history of fever for 6 days, along with cough and pronounced fatigue; temperature: 37.5–38.6°C. The patient was residing in Wuhan on January 14, 2020, and had previous contact with confirmed patients. Laboratory tests: PLT 1.4×109/L ↓， CRP 7.06 mg/L ↑， EO% 0.2 ↓， EO 0.01 ↓， AST 36.40 ↑， LDH 289 ↑. CT scans showed multiple patch GGOs in both lungs and fibrosis lesions in the lower lobes.

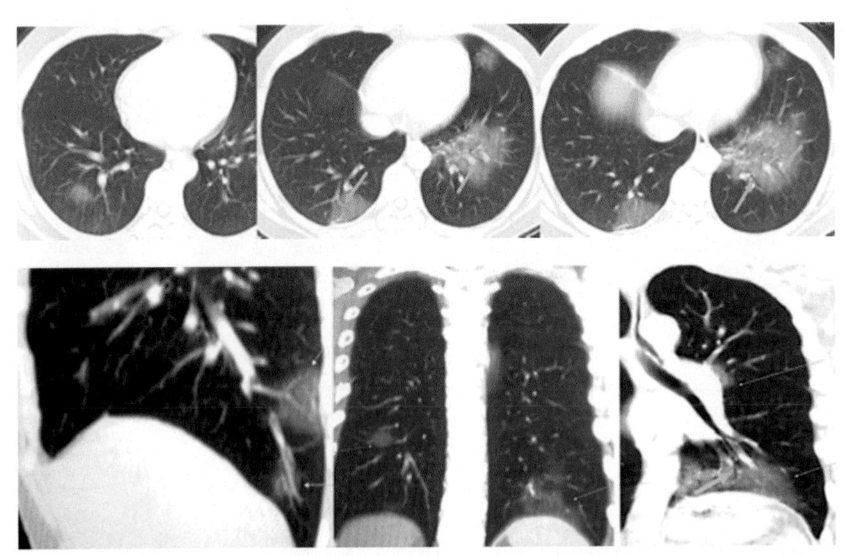

Fig. 3.17 Male, 41 years old, with a history of fever and fatigue for 10 days; temperature: 38.3°C; with a history of contact with confirmed patients in Wuhan 10 days previously. Laboratory tests: WBC 3.71×10^9/L， LYMPH 0.88×10^9/L， ESR 15 mm/h； CRP 13.50 mg/L， hs-CRP > 3.00 mg/L. HRCT of progressive stage showed multiple, massive, and patch GGOs with vascular enlargement and bronchiolar dilatation in both lungs.

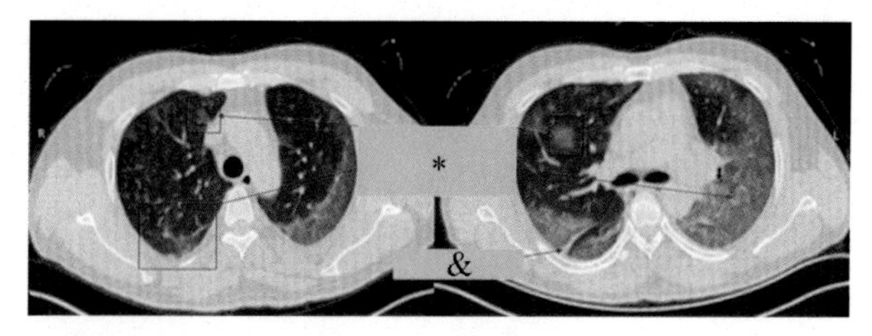

* Bilateral, multiple, and massive patch GGOs

(A) &Thickening oblique fissure of the right lung

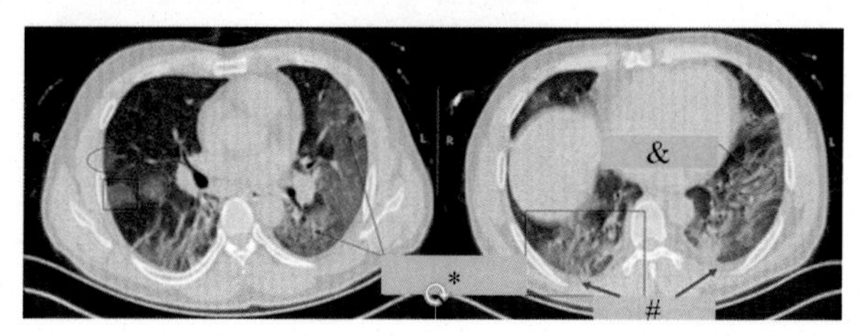

*GGOs with small vessels enlargement

&Air bronchogram sign

(B) #a small amount of pleural effusion

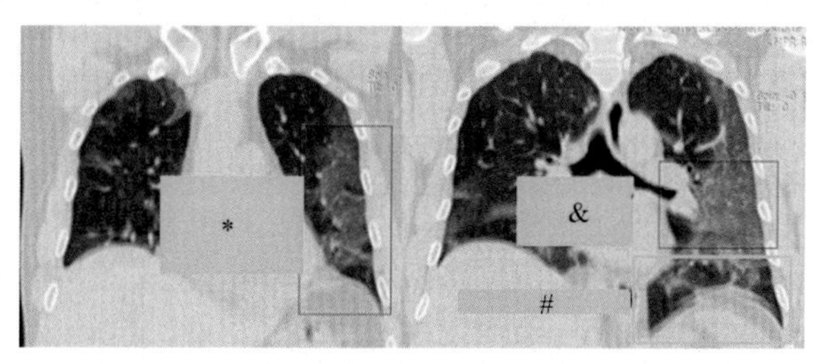

*Air trapping, which is due to the coexistence of exudation lesions and small airway changes, results in mosaic sign.

&GGOs with the increased number of vascular and vascular enlargement

(C) #There is air trapping in the distal area of lesions.

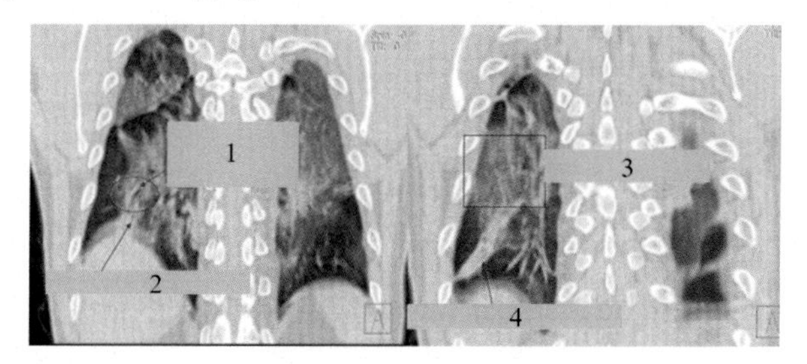

1 Thickening bronchovascular bundles and air bronchogram sign

2 Localized subsegmental atelectasis

3 Increased number of pulmonary small vascular networks

(D) 4 Localized subsegmental atelectasis

Fig. 3.18 (***See figure legend on next page***)

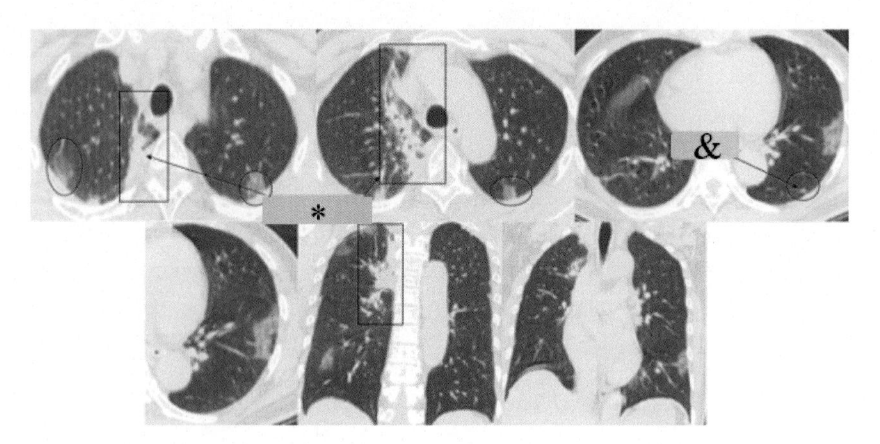

*Radiation pneumonia

&Small pulmonary nodules

Fig. 3.19 Male, 46 years old, with a history of fever for 4 days, along with paroxysmal cough; temperature: 37.7°C. He was residing in Wuhan on January 20. Laboratory tests: WBC 2.99×10^9/L, LYMPH% 16.7%. He had a history of adenocarcinoma in the right upper lobe and was treated with chemoradiotherapy. CT scans on January 25 showed multiple small patch GGOs in both lungs and bronchiolar dilatation in large lesions. There was radiation pneumonia in the right upper lobe, and the tumor was stable.

Fig. 3.18, cont'd Male, 50 years old, with a history of fever for 4 days, along with dry cough, fatigue, and chest distress; temperature: 38.2°C. The patient returned from Wuhan by train on January 16, and he was involved in clusters of COVID-19 cases. (Left column) CT scans on January 23 showed GGOs with air bronchogram and vascular enlargement in the right upper lobe. (Right column) Increased number of and enlarged lesions were observed in the right lung on January 26. Meanwhile, new lesions appeared in the left lung.

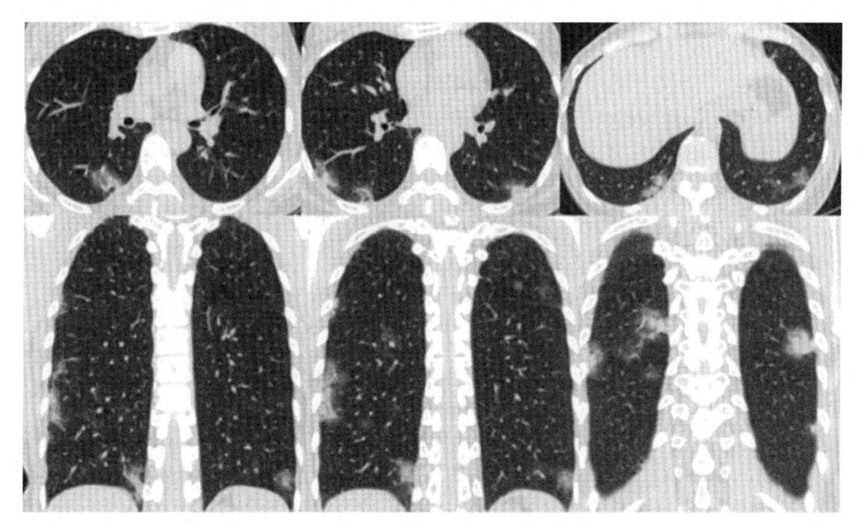

Fig. 3.20 Female, 45 years old, with a history of fever for 4 days, along with fatigue and paroxysmal cough; temperature: 37.6°C. She and the patient in Fig. 3.19 are a couple. She returned from Wuhan on January 20, 2020. Laboratory tests: WBC 4.24×10^9/L, LYMPH% 26.25%. The patient had positive results on RT-PCR of pharyngeal swab and blood specimens on January 24, 2020. CT scans on January 25, 2020 showed multiple small patch GGOs in the lower lobes and the left upper lobe. All lesions were located in the lung periphery and some were accompanied by vascular enlargement.

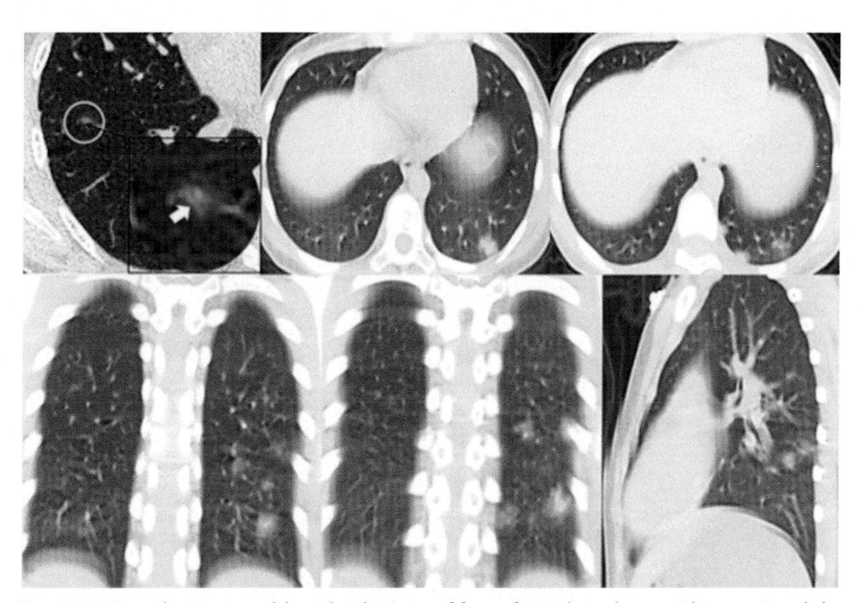

Fig. 3.21 Female, 9 years old, with a history of fever for 1 day, along with occasional dry cough but no sputum; temperature: 38°C. She is the daughter of the patients in Figs. 3.19 and 3.20. She returned from Wuhan with her parents in the same car on January 20. Laboratory tests: WBC 5.70×10^9/L， LYMPH% 41.80%， LYMPH 1.87×10^9/L. The patient had positive results on RT-PCR of pharyngeal swab and blood specimens January 24, 2020. CT scans on January 26, 2020 showed multiple, small, and circle GGOs in the lower lobes.

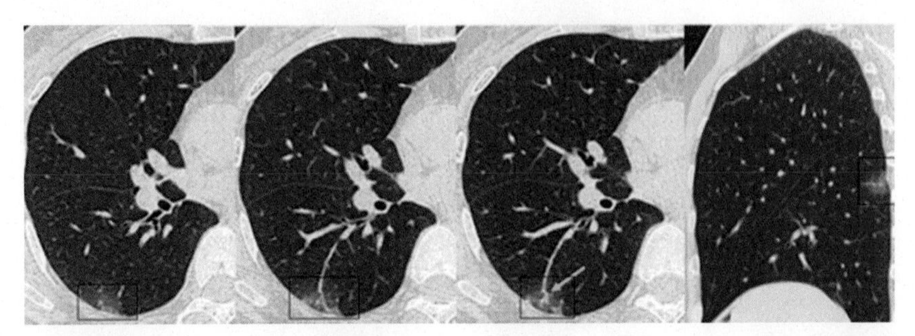

Fig. 3.22 Female, 51 years old, with a history of fever for 2 days and paroxysmal cough for 1 day; temperature: 38°C. She is the sister of the patient in Fig. 3.19 and was residing in Wuhan. Laboratory tests: WBC 3.82×10^9/L, LYMPH% 23.3%, NEUT% 81.51%. CT scans on January 26, 2020 showed GGOs with vascular enlargement in the dorsal segment of the right lower lobe.

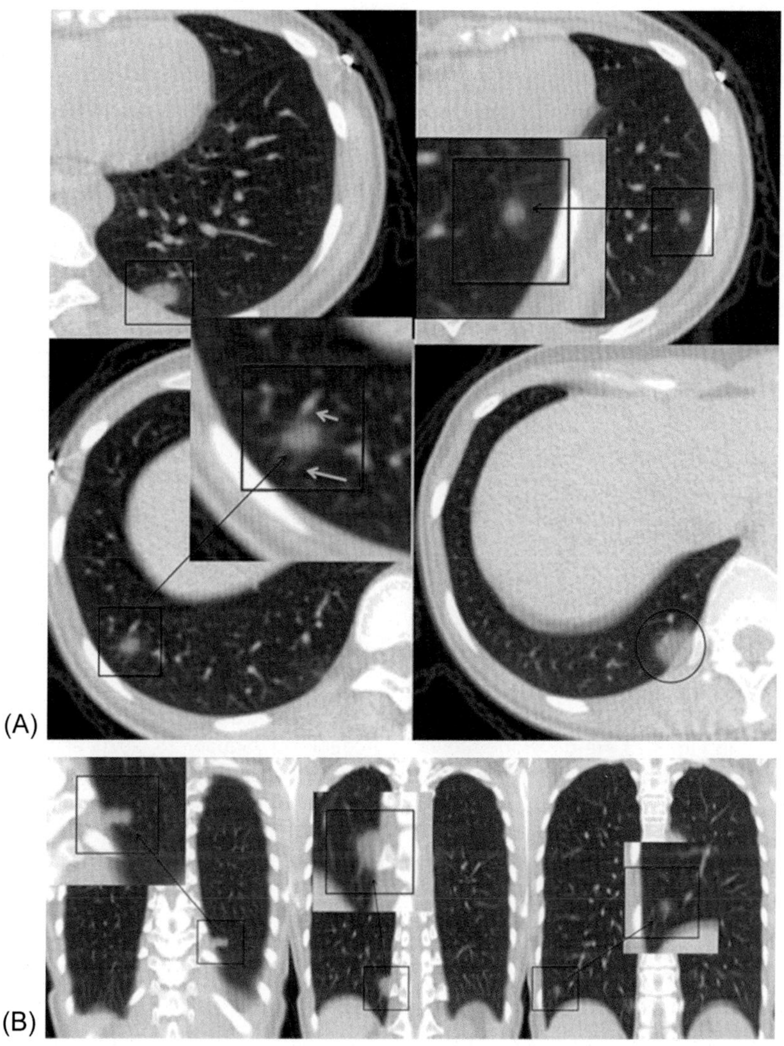

Fig. 3.23 A 32-year-old female patient was residing in Wuhan and returned to Xi'an on January 21, 2020. She visited hospital without fever on January 27, 2020. Laboratory tests showed she had a positive result on RT-PCR, while WBC and N% were normal. (A) CT scans showed solid nodules of varied size. Some lesions were accompanied by halo signs and others were with smooth margins. (B) There were solid nodules with a subpleural distribution in the posterior segment of the lower lobes. Vascular enlargement was observed in the periphery of the right lower lobe, surrounded by exudative lesions.

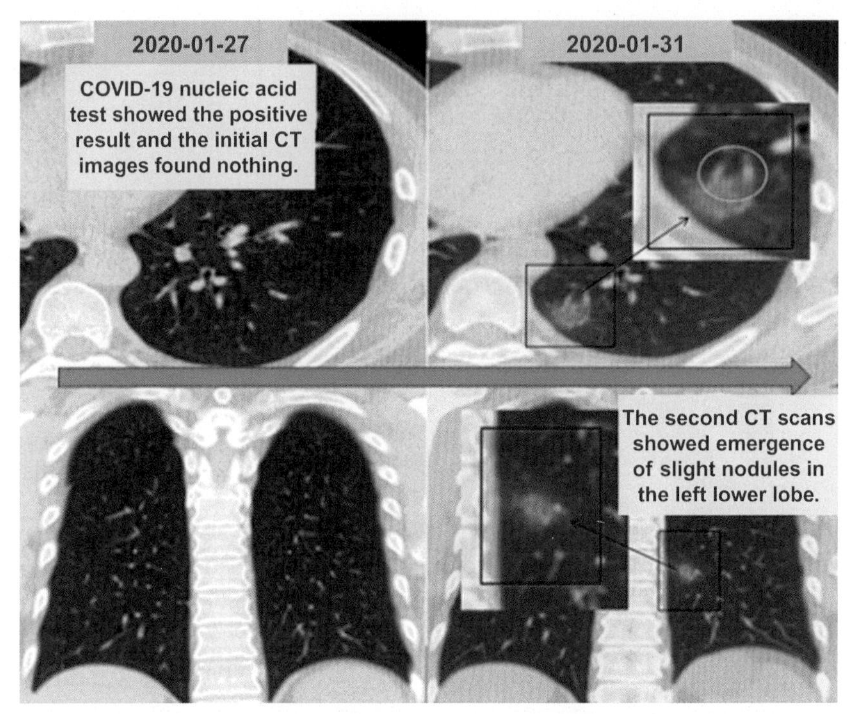

Fig. 3.24 A 44-year-old male patient who was residing in Xi'an and had contact with the patient in Fig. 3.23. He developed fever and cough on January 26, 2020, and visited hospital the next day. Laboratory tests showed he had a positive result on RT-PCR, while WBC and N% were normal. (Left column) Initial CT scans showed there were no evident exudative lesions or consolidation. (Right column) Follow-up CT scans on January 31 showed emergence of new slight nodules in the left lower lobe.

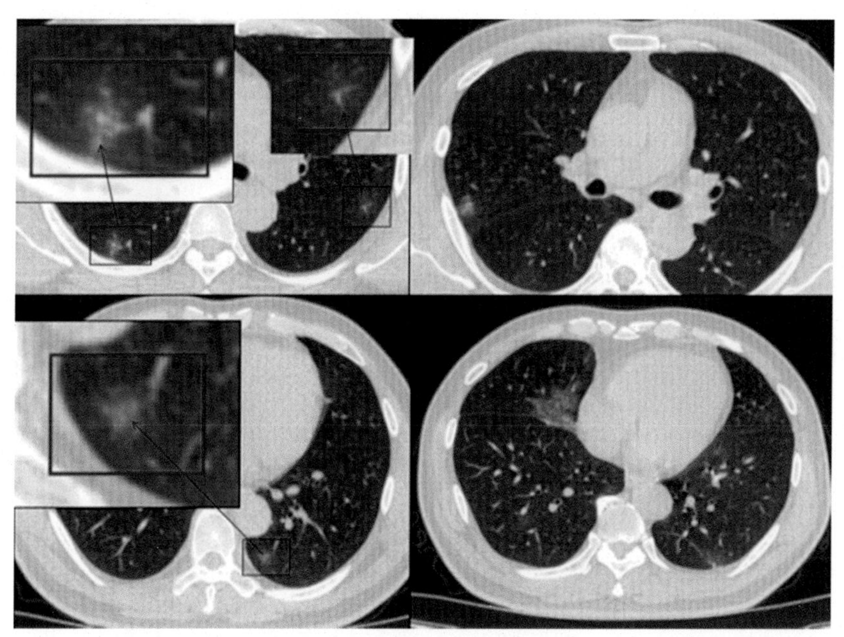

Fig. 3.25 A 47-year-old man who developed fever, cough, expectoration, abdominal distention, and poor appetite on January 23, 2020. He visited hospital on January 27, 2020. Laboratory tests showed he had a positive result on RT-PCR, while WBC and N % were normal. CT scans showed multiple and mix GGOs of varied size with obscure margins in the subpleural area. The majority of lesions were quite small, but the lesions in the medial segment of the right middle lobe were massive.

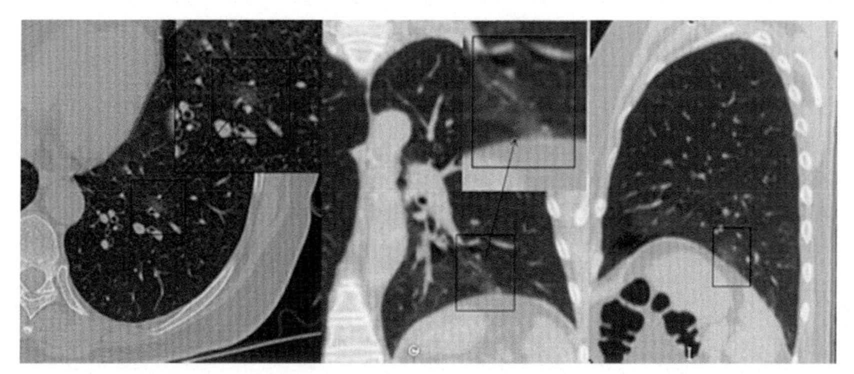

Fig. 3.26 A 47-year-old female patient, who is the wife of the patient in Fig. 3.25, developed fever, cough, expectoration, abdominal distention, and poor appetite on January 27, 2020. She visited to hospital on January 29, 2020. Laboratory tests showed she had a positive result on RT-PCR, while WBC and N% were normal. CT scans showed a small amount of GGOs in the left lower lobe and along the bronchovascular bundles.

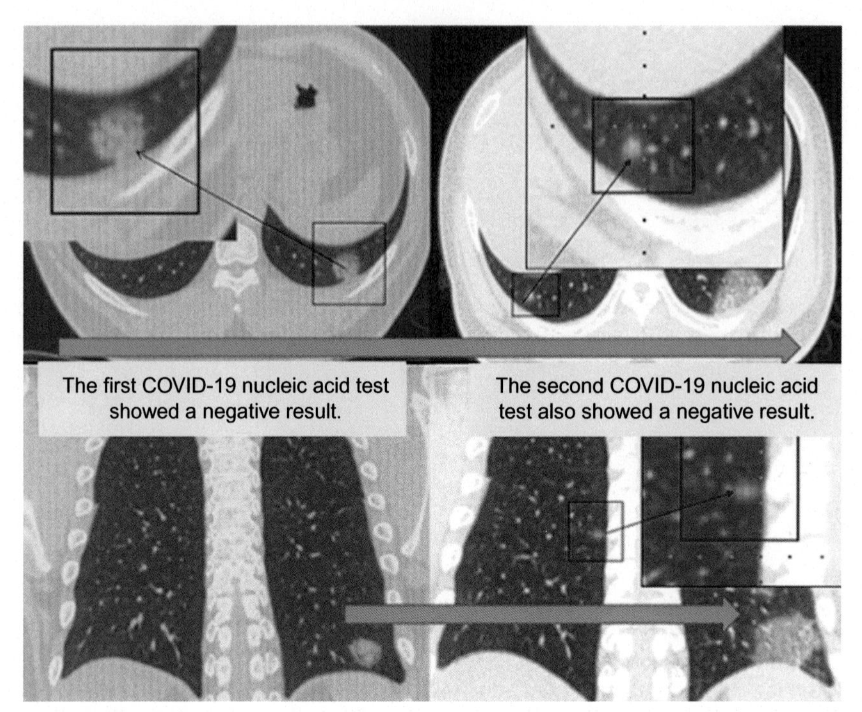

Fig. 3.27 A 22-year-old man, who is the son of the patients in Figs. 3.25 and 3.26, developed fever on January 27, 2020. Nucleic acid detection was performed twice and had negative results. Laboratory tests showed WBC 11.75, NEUT 9.38, N% 79.8%, LYM 1.13, LYM% 9.6%. CT scans showed patch GGOs with obscure margins and air bronchogram signs in the left lower lobe. Follow-up CT scans showed enlarged lesions in the left lower lobe and emergence of new exudation lesions.

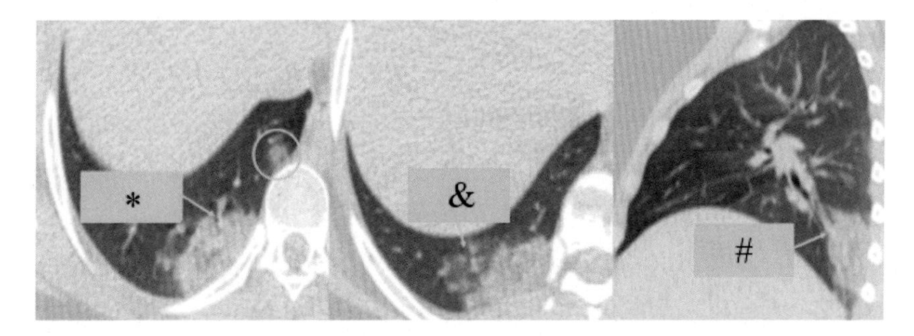

(A)
*Lesions with vascular enlargement
&Lesions surrounded by GGOs
#Lesions with vascular enlargement

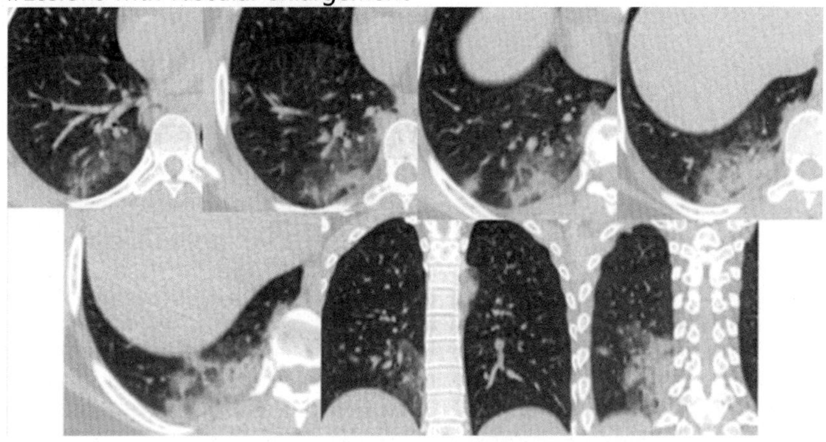

Fig. 3.28 Female, 22 years old, with a history of fever for 3 days, but no cough, expectoration, sore throat, or other symptoms; temperature: 38.1°C. She drove to Wuhan and stayed there for 9h on January 18. Laboratory tests: WBC 6.5×10^9/L，LYMPH% 18.6%. (A) CT scans on January 27, 2020 showed massive mix-pattern consolidation surrounded by GGOs in the right lower lobe. (B) Follow-up CT scans showed new, multiple, and small patch exudative lesions in the right lower lobe. Additionally, coexistence of absorption of primary lesions and emergence of enlarged ones were observed in the right lower lobe.

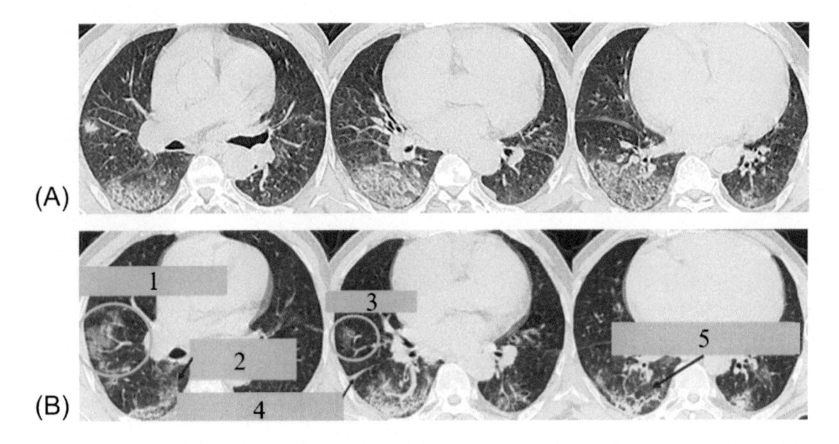

1 Increased number of or enlarged lesions

2 Primary lesions were partially absorbed and changed into fibrosis.

3 New lesions

4 Local thickness and displacement of interlobular fissure

5 Fibrosis lesions and traction bronchiectasis

Fig. 3.29 Male, 60 years old, with a history of fever for 1 day and fatigue for 5 days; temperature: 38.5°C. He had returned home from Wuhan 5 days previously. Laboratory tests: CRP 43.15 mg/L, hs-CRP > 5.00 mg/L, ESR 15 mm/h. (A) CT scans on January 23, 2020 showed multiple, bilateral, and patch consolidation lesions of varied size surrounded by GGOs. There were reticular patterns and air bronchogram signs inside lesions. (B) Follow-up CT scans on January 28, 2020 showed lesions in the right lower lobe were partially absorbed and changed into fibrosis. There were new, bilateral, and patch GGOs and consolidation lesions with a peripheral lung and subpleural distribution. Compared with the CT images of 5 days before, we found the location of horizontal fissure was different.

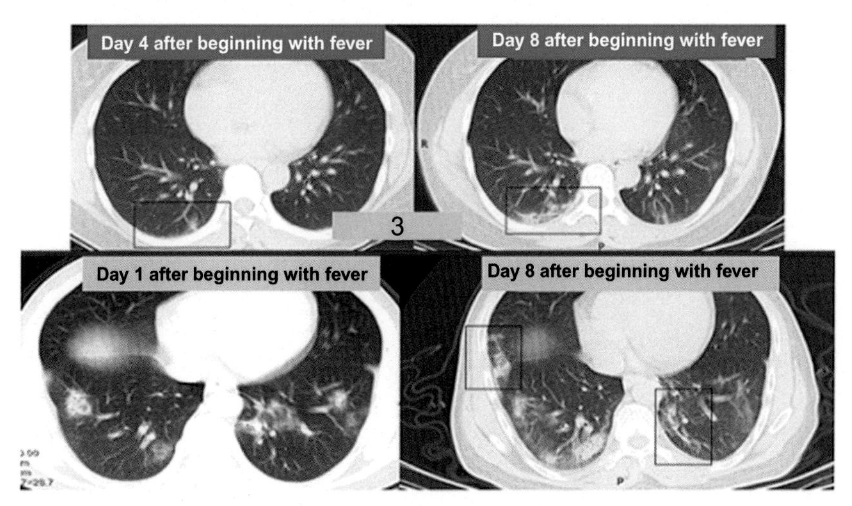

Fig. 3.30 Female, 64 years old, with a history of fever for 6 days, along with cough. She lived in Wuhan and was a community medical worker. Initial CT images indicated slightly thickening lung markings. She was treated for upper respiratory tract virus infection without improvement. The temperature fluctuated between 37.5°C and 38.6°C and she felt pronounced fatigue. She developed a more severe cough and occasional dyspnea 3 days later. Follow-up CT scans on day 6 after beginning with fever showed that the shape and range of lesions changed greatly. In addition, fibrosis lesions were observed in the subpleural area of bilateral lower lobes.

CHAPTER 4

Diagnosis of COVID-19

Contents

1. Diagnosis

The diagnosis of COVID-19 is based on epidemiological history, clinical manifestations, and pathogenic confirmation.

1.1 Definition of suspected case

Patients who meet any one of the epidemiological criteria and any two of the clinical manifestations are suspected to have COVID-19. If there is no clear epidemiological history, three of the clinical manifestations should be met.

Since the first edition of the pneumonia diagnosis and treatment plan for novel coronavirus infection (trial) issued by the National Health Commission, it has been updated to the seventh edition, most of which was revised from late January to mid-February (second edition on January 18, third edition on January 22, fourth edition on January 27, fifth edition on February 4, revised on February 8, and sixth edition on February 18). The consecutive update of guidelines suggests that diagnosis should be constantly revised based on up-to-date clinical experience. The change of the definition of epidemiology is a good example. The current epidemiological history is as

COVID-19
https://doi.org/10.1016/B978-0-12-824003-8.00004-8

follows: (1) history of residence in or travel to Wuhan and surrounding areas or other communities with positive case reports within 14 days before onset; (2) history of contact with SARS-CoV-2-infected persons (with positive nucleic acid test) within 14 days before onset; (3) history of contact with patients from Wuhan and surrounding areas, or patients with fever or respiratory symptoms from a case-reporting community within 14 days before the onset of illness; or (4) clustered cases. Compared with the original diagnosis and treatment plan, the main change in the history of epidemiology is the update of travel history or residential history to Wuhan and surrounding areas, or other case-reporting communities. The first reason is that other areas have also found community cases, and second, the "exposure to fever and respiratory symptoms" has become increasingly blurred. It has also been emphasized that there is a history of contact with SARS-CoV-2 infected persons, and SARS-CoV-2 infection refers to those with positive detects of pathogenic nucleic acids, no matter where they come from.

At the same time, the definition of clinical performance has also been revised because of a better understanding of the disease. On the other hand, a contemporary revision fulfills the needs of different regions. For example, the primary update in clinical performance in the fifth version is to distinguish Hubei from other provinces. The diagnostic criteria for suspected cases in Hubei were greatly extended. In the diagnostic criteria of suspected cases (Hubei), the clinical manifestations were: fever and/or respiratory symptoms; the total leukocytes count in the early stage was normal or decreased; or reduced lymphocyte count. Compared with the previous clinical manifestations, it added "and/or respiratory symptoms" and did not require pneumonia. Patients with no definite history of epidemiology and those who met two of the clinical symptoms simultaneously were diagnosed as suspected cases. In the definition of suspected cases in other provinces except Hubei, the clinical manifestations were: fever and/or respiratory symptoms; imaging features of pneumonia; normal or decreased white blood cells at the early stage of onset; or decreased lymphocyte count. In accordance with any two of the clinical manifestations, regarding "without pneumonia" patients from outside of Hubei, pneumonia manifestations are not required in order to avoid missed diagnosis. Any of the three clinical manifestations are required with no specific epidemiological history. Starting from the sixth version, a difference in origin is no longer included in the definition of clinical manifestation, unified as follows: (1) fever and/or respiratory symptoms; (2) consistent with the imaging of COVID-19; or (3) normal or decreased total count of white blood cells in the early stage of the disease and normal or decreased lymphocyte count.

1.2 Definition of confirmed case

If any one of the following pathogenic or serological tests is positive, the patient is confirmed as COVID-19: (1) positive RT-PCR results for SARS-CoV-2 nucleic acid; (2) viral gene sequencing highly homologous to the known SARS-CoV-2; or (3) serum samples positive for SARS-CoV-2-specific IgM and IgG antibodies. The SARS-CoV-2-specific IgG antibody is required to change from negative to positive or the titers in the recovery period are required to be four times or higher than those in the acute phase.

In the fifth version of the plan, blood samples were added to the specimens except for respiratory tract specimens. Blood samples were collected from patients from whom it was difficult to obtain respiratory secretions. The sixth version was further revised, and nucleic acid tests no longer emphasized the detection of samples. As long as the SARS-CoV-2 was positive, it could be counted as a confirmed case.

1.3 Clinical diagnosis

The fifth edition of the program was specially designed for Hubei to establish the diagnostic criteria of "clinical diagnosis cases," which include clinical compliance with the characteristics of viral pneumonia, such as corresponding clinical symptoms and imaging CT findings, especially the multiple lobes exudative ground-glass shadow and intermittent consolidation, normal or decreased total count of white blood cells in laboratory examination, and reduced lymphocyte count. Even if there is a lack of nucleic acid test results, it can also be applied for clinical diagnosis. For patients who meet the clinical diagnostic criteria of COVID-19, they should be treated within a restricted space.

The reason for these criteria is that Hubei and Wuhan city have gathered a large number of suspected patients waiting for nucleic acid testing. According to the nucleic acid detection capability of Hubei province, it is difficult to complete all tests in a short time. Based on the intention of early diagnosis and treatment, such a clinical standard should be applied in Hubei in order to ensure that all patients receive timely treatment without developing into a severe or critical condition, in line with clinical practical needs. Meanwhile, patients with a clinical diagnosis should be isolated in time to reduce the risk of infection transmission. This is not applicable nationwide because cases throughout the country are sporadic, so we have adequate time for diagnosis and etiological examination. It is possible to give priority to frontline isolation treatment. Therefore, the relevant provisions of the fifth edition are applicable only to Hubei province. Subsequently, this province has rapidly admitted suspected cases within a short period of time, and its nucleic acid detection capability has become able to meet the clinical needs. There is no requirement to distinguish between inside and outside Hubei province. In this context, when the sixth version of the plan was introduced, the clinical diagnosis criteria were unified in and out of Hubei province.

2. Etiological test

2.1 Respiratory tract virus detection

Novel coronavirus pneumonia is actually a form of community-acquired pneumonia (CAP). Pathogens causing CAP include bacteria, viruses, and atypical pathogens. In terms of when and what specimens should be collected for laboratory tests among CAP patients, the guidelines for diagnosis and treatment of adult community-acquired pneumonia in China (2016 edition) clearly recommend that under normal circumstances, outpatients do not have to receive an etiological examination. However, laboratory tests should be carried out under certain circumstances and for hospitalized patients. Etiological examinations are highly recommended in specific clinical situations with CAP. Five situations require respiratory virus screening: cluster outbreak, initial empiric treatment failure, severe cases, multiple lobe lesions, and immunodeficiency.

In cases where viral pneumonia is strongly suspected, the relevant virological examination should be considered. Currently, there are four recommended assays, including virus isolation and culture, serum-specific antibody detection, virus antigen detection, and nucleic acid detection. Since virus isolation and culture are very challenging, they are only used for laboratory research instead of as diagnostic criteria. Serum-specific antibody tests include IgM and IgG. IgM presents only within a shorter time, while IgG is only found in convalescent serum, which limits its value for diagnosis. Therefore, in practical clinical diagnosis, there are two main indicators of viral pneumonia. One is viral antigen detection, which includes direct immunofluorescence assay (DFA) and the colloidal gold method. The colloidal gold method is more widely used; however, although it has good specificity, its sensitivity is not high. The other indicator is nucleic acid detection. In the diagnosis of influenza A, for instance, the main methods of clinical application are virus antigen detection (such as colloidal gold detection) and nucleic acid detection. Therefore, nucleic acid detection is critical.

2.2 Nucleic acid detection for SARS-CoV-2

In the detection of SARS-CoV-2, nucleic acid detection is used as the laboratory diagnostic standard of confirmed cases. Positive means that novel coronavirus nucleic acids can be detected in specimens of nasopharyngeal swabs, sputum, lower respiratory tract secretions, blood, or feces. In the sixth edition, the source of samples is no longer required. As long as nucleic acid detection is positive, it can be diagnosed. The methods are: (1) real-time fluorescence RT-PCR detection of SARS-CoV-2 nucleic acid positive and (2) viral gene sequencing, highly homologous with the known novel coronavirus.

Nucleic acid detection is an irreplaceable diagnostic method: it is critical for optimized use of limited medical resources and timely treatment of patients. First, the detection of pathogen antibodies has a window period, and antibodies present at the late phase, usually later than nucleic acid detection. Second, the sensitivity of nucleic acid detection is much higher than other laboratory detection methods, as mentioned above. Third, nucleic acid detection is possible earlier than detection through changes in chest CT images, which has allowed us to identify asymptomatic COVID-19 patients. Fourth, quantitative nucleic acid detection can enable us to monitor dynamically the level of virus infection and observe the therapeutic effect.

PCR detection is a method to detect the change of amplification product from each PCR cycle by measuring fluorescence intensity, which is a quantitative analysis method for the initial template. Many clinical real-time quantitative PCR detection kits are available for clinical use. Although PCR tests are very sensitive, the detection rate can only reach 30%–50%. It requires clinicians to go carefully over the epidemiology, clinical imaging, and dynamic changes, and not simply rule out infection due to a negative result of an upper respiratory tract examination. Dynamic follow-ups and repeated examinations can help improve the accuracy of diagnosis.

The wide application of the next generation of gene sequencing technology (NGS) has given us a better understanding of the sequencing of the pathogenic genes. NGS is designed to sequence the gene of the specimen. If the gene is highly homologous with the known SARS-CoV-2 gene, then it can be determined. The method is accurate and effective, which can be used in final confirmation for a controversial sample, monitoring of virus mutation, and origin tracing, but the cost is relatively high. The process is complex, requiring sequencing and professional bioinformatics analysis.

Gene sequencing offers various advantages. First, the virus cannot be detected by PCR when mutation happens during the transmission process. However, high depth metagenomic sequencing (the total reads number of sequencing is no less than the 80 M DNA sequence) can compensate for the shortcomings of RT-PCR and monitor the potential variation. Second, viral load in some samples is too low to be detected by RT-PCR. Metagenomic sequencing assay can effectively improve the detection rate when the human origin background is low. Third, for suspected cases with negative detection of SARS-CoV-2, metagenomic sequencing can provide effective information on other possible pathogens. Fourth, metagenomic sequencing can also provide related pathogen information on multiple or secondary infection.

The PCR assay is easy to operate and fast with low cost, making it more suitable for processing large numbers of patients, whereas NGS is expensive and time consuming, which is difficult for mass inspection. Therefore, PCR is more applicable for census. In cases where a patient happens to have a positive result followed by a negative one, a gene-sequencing test should be used. If a patient has typical clinical symptoms but provides a negative result, we recommend the use of NGS because existing PCR kits may fail in detection due to gene mutation. Moreover, in critically ill patients who are suspected of mixed viral and bacterial infections besides SARS-CoV-2, or with immunocompromised status such as diabetes or receiving immunosuppressive agents, NGS is also strongly recommended. We have to make use of NGS because it has more advantages than PCR assay.

A combination of RT-PCR and metagenomic detection is highly recommended to detect SARS-CoV-2 comprehensively and effectively, as well as to monitor the potential viral mutation during the transmission process.

2.3 Problems and troubleshooting in PCR testing

At present, the problems encountered in virus nucleic acid detection may come from the following:

(1) Timing and location malfunction of specimen collection. The organs impacted by the virus change over time. The virus affects the upper respiratory tract at the onset of the disease. This then clears, while the viral load in the lower respiratory tract increases. Therefore, it would be better to collect specimens from the lower respiratory tract (deep sputum, BALF). BALF is optimal but it is difficult to obtain and time consuming, with a high risk of transmission. Not all patients require this or are suitable. Among critically ill patients, deep sputum is also not easy to obtain, but it would be more useful, even if it is"saliva sputum," than specimens of nasopharyngeal swabs. Although nasopharyngeal swabs are most commonly used, their optimal collection time remains unknown. According to the characteristic of influenza A, it reaches a peak at 24–72h and then drops rapidly. It is possible that SARS-CoV-2 may have a similar rate of change to influenza and become undetectable from nasopharyngeal swabs at the late stage. In addition, limitation of accuracy may also be related to human factors. The firstline officials undergo enormous risk of infection, which may lead to incomplete and malfunctional sampling. The standards of procedure as well as the quality of swab product are also crucial.

(2) Sample shipping limitation. The preservation and transportation of samples have influence on the results. Unlike DNA virus, RNA virus is much more likely to mutate and degrade. During the collection process, the nucleic acid of RNA virus is one of the most difficult biological molecules to maintain in a stable condition due to self-degradation and enzyme-mediated degradation. Samples containing viral RNA should be stored in a dedicated virus preservation solution, and must be stored in refrigerated conditions (i.e., 4°C or colder) and delivered for inspection as soon as possible. Whether a laboratory has the capability to conduct a professional test is also one of the key factors.

(3) Reagent kit issue. Several commercial PCR kits were introduced after the nucleic acid detection standard was announced. The national FDA initially screened 7 of 53 capable companies, and 4 of them eventually passed the expedited process on January 26. They were the detection kit for SARS-CoV-2 from Beijing Genomics Institution (BGI), DNBSEQ-T7 sequencing system, SARS-CoV-2 nucleic acid detection kit from Shanghai Jienuo (RT-PCR), and SARS-CoV-2 nucleic acid detection kit from Shanghai ZJ Bio-Tech (RT-PCR), followed by another three companies. However, such a rush launch inevitably brought much negative feedback regarding product quality and overpromising in advertisements, such as "quick response in only ten minutes," and "100% sensitivity and specificity." In addition, more clinical evidence is still required as a reference upon laws for registration renewal as well as subsequently clinical application.

(4) The absence of large sample studies and clinical evidence. It needs to be determined whether these detection kits detect completely and accurately in clinical application. Most of the novel coronavirus reagents are targeted at specific areas of SARS-CoV-2. The *ORF1ab* gene and *N* gene are measured with fluorescence intensity by RT-PCR. The current reagent can only design primers based on existing data, which limits comprehensive understanding in the real world. In addition, the research on SARS-CoV-2 has just started; several questions need to be answered, such as the frequency of variation of viral gene, mutation hotspots, or conservative regions of evolution. Whether any false negatives exist due to the mutation in the amplification area requires further observation.

Future strategies and development directions for nucleic acid detection are as follows:

(1) Standardize the protocol of specimen collection. Try to collect lower respiratory tract specimens, especially BALF or deep phlegm, to improve the positive rate. Master timing of collection: nasopharyngeal swabs for the early phase and BALF for the late phase. However, this does not mean that nasopharyngeal swabs are useless at the late stage. For instance, two patients with confirmed COVID-19 were hospitalized for almost 20 days. During hospitalization, nucleic acid testing was conducted intermittently and the results remained positive until day 20. Standard operation procedures of nasopharyngeal and throat swabs include appropriate personal protective equipment (medical N95 masks, goggles, gloves, protective masks, protective suits, etc.), authorized nasopharyngeal and throat swab products for collection, and standard laboratory operating procedures. Unlike ordinary cotton swabs, well-designed swabs are expensive. Samples can be collected through the pharynx as well as the nasal cavity. If we only collect in the area of the nostrils, the positive rate will be low. Therefore, the method of collecting nasopharyngeal swabs is critical.

Regarding the sampling sites, nucleic acid can be collected from five places: (1) nasopharyngeal swabs; (2) saliva under the tongue when getting up next morning, where the virus accumulates easily; (3) deep respiratory secretions, including phlegm, aspiration, or bronchoalveolar lavage fluid; (4) blood, in which the viral load is relatively high during viremia; and (5) anal swabs, preferring rectal mucosa swabs. Based on experience from some institutions, the anal swab can still be positive when the throat swab becomes negative. Faced with this novel disease, we have to avoid deducing the whole picture of its pathogenesis only from partial experience. We have to think of the whole procedure from sampling to detection, and take into account collection sites as well as distribution of the virus *in vivo*.

(2) Improve the delivery of specimens. Collect and ship specimens in time; reduce the degradation of RNA virus nucleic acid; store in a dedicated viral preservation solution and refrigerator (i.e., 4°C or lower); and expedite the transportation process and save waiting time for detection.

(3) Improve approval process of kit production. To expedite the verification and approval process of reliable reagent manufacturers, the National Health Commission has verified seven companies that can produce reliable commercial detection kits for clinical application. Meanwhile, we have to promote the construction of a P2 laboratory (Physical Containment Level 2) in the designated hospital, to give access to more hospitals. More attention needs to be paid to the biosafety issues in laboratories. In order to ensure the quality of test reagents, all manufacturers need to complete all the clinical verifications in accordance with the requirements of laws and regulations so that they can be registered for routine clinical diagnosis.

The performance from different manufacturers is not equal. The reasons are as follows: ① The specificity and efficiency of primer probe systems require good reference sequences, bioinformatics analysis ability, and suitable target areas. ② The stability of the process system requires extensive development of *in vitro* diagnostic reagents and the establishment of a stable and efficient production process system. We hope to select reliable kits with higher sensitivity to improve the positive rate and reliability. We should select the reagents with stable performance that are validated by clinical positive samples and subsequently obtain NMPA registration certificates. In this way, the quality of products can be guaranteed.

(4) Improve the detection technology. A standard virus-specific detection area is required. SARS-CoV-2 is a linear single-stranded RNA (ssRNA) virus with a total length of 29,903 nucleotides, containing 10 genes. After analyzing the complete genome sequence of SARS-CoV-2, three genetic regions (ORF1ab, E, N) of the virus were recommended by the National Health Committee, which can be used as target sequences for the design of primer probes. Coronavirus is a large class of viruses, including SARS, MERS, and other coronaviruses causing the common cold. The E and N genes are relatively conserved in the coronavirus, especially within coronavirus 229E/OC43/HKU-1, which can infect human beings but has poor transmission capability. The virus above exists within the natural environment cycle; therefore, target sequence detection may lead to cross positive responses from other species of coronavirus. To avoid cross contamination, a good reference sequence and powerful bioinformatics analysis capability are required to design the specific region of novel coronavirus

for suitable primer probes. Recent studies showed that the specific region mainly concentrates on the *ORF1ab* gene and *S* gene. For the detection genes, fewer are better. In fact, the primer probes will interfere with each other, leading to the decrease of sensitivity. The double domain will be weaker than the single one, and triple will be still weaker. For example, in HIV detection kits, the target sequence includes *gag* and *env* genes, and only one of them is detected, which has already achieved high sensitivity and accuracy.

(5) Comprehensive assessment (back to bedside). Finally, we have to emphasize that results of nucleic acid tests should be interpreted back to clinical practice. We should speed up the production of kits of good quality, improve construction of laboratories, and standardize clinical practice, especially in primary hospitals and healthcare centers. We should standardize the process of collecting clinical specimens and laboratory testing to ensure the reliable interpretation of experimental results. Based on the clinical manifestations and epidemiological history, it would be better to make a comprehensive evaluation and to avoid making a decision only on numerical results.

In conclusion, the optimal detection methods need selection from the baseline. First of all is quality control of the sampling process, including whether the swab and the transport medium used for sampling, sampling manipulation, and storage and shipping temperature are appropriate. Second, a reliable detection kit is required. Finally, a reasonable judgment and interpretation of the results are also required. Since various pathogens can share similar signs and symptoms of respiratory infection, we should not ignore the possibility of other respiratory tract pathogens.

2.4 Antibody detection assay for SARS-CoV-2

To date, there are few antigen detection reagents for clinical use. The detection of coronavirus antigen is prone to cross reactions. It is difficult to select antigen recognition sites and the key point is to find the specific designed binding domain for antibodies. For example, the antigen detection reagent developed by the Institute of Pathogen Biology of the Chinese Academy of Medical Sciences/Beijing Union Medical College is facing a challenge in how to select more sensitive antibodies by ensuring the level of specificity. The current reagents need to solve the cross contamination issue with known pathogens, and need to be refined constantly. Antigen test assay is characterized by rapid and specific diagnosis—for instance, that used in influenza viruses detection. We hope that the method of antigen detection assay can be applied to COVID-19 detection in the near future.

Antibody detection assay has become a relatively reliable method in clinical use. There are many kinds of antibodies, such as viral-specific antibodies, IgM at the early phase, and IgG at the convalescent phase, which is a major component of neutralizing antibodies. We should also pay attention to IgA at the early phase, which presents at the same time as IgM or even earlier. However, the role of airway-secreted IgA requires further investigation.

In addition, the evaluation of neutralizing antibodies determines the real antiviral effect; however, to date, it has not become feasible to evaluate neutralizing antibodies. The RBD antibody detection assay seems to have only a weak prediction value on neutralizing antibodies, but more assays are needed. Currently, specific IgG and IgM are mainly used for disease diagnosis. If we plan to use convalescent plasma as a therapeutic strategy, we have to preevaluate the level of antibodies in the donators, at least the level of IgG. The evidence suggested that the plasma level of specific IgG at the recovery stage corresponds to the level of the neutralizing antibody. In a recent study from Wuhan Tongji Hospital, two critically ill patients received convalescent plasma and the level of neutralizing antibody reached up to 1:640. In short, antibody detection is not only for pathogen diagnosis but also for the overall situation of the disease.

3. Severity evaluation

COVID-19 can be classified as mild, moderate, severe, and critical, as follows:

(i) *Mild:* Clinical symptoms are few, with no pneumonia manifestation on lung imaging. (ii) *Moderate:* Fever and respiratory symptoms with pneumonia manifestation on imaging; without dyspnea or other complications. (iii) *Severe:* Patients meet any of the following criteria: shortness of breath, respiratory rate (RR) $\geq$30 beats/min; resting state, oxygen saturation $\leq$93%; partial pressure of arterial oxygen (PaO_2)/fraction of inspired oxygen (FiO_2) $\leq$300 mmHg (1 mmHg = 0.133 kPa). At higher altitudes (>1000 m), PaO_2/FiO_2 should be corrected according to the following formula: $PaO_2/FiO_2 \times$ [atmospheric pressure (mmHg)/760]; pulmonary imaging shows that the lesions have developed by >50% within 24–48 h. (iv) *Critical: Patients with one of the following criteria:* respiratory failure requiring mechanical ventilation; shock; and multiorgan failure requiring ICU monitoring and treatment.

We list the following risk factors of severe COVID-19: (i) elderly patient (age >65 years); (ii) comorbidities, such as hypertension, diabetes, and coronary heart disease; (iii) progressive decrease in peripheral blood lymphocytes, CD4+ T lymphocyte count $<250/\mu L$; (iv) progressive increase in serum level of inflammatory factors, such as interleukin (IL)-6 and C-reactive protein; (v) progressive increase in lactic acid and lactic dehydrogenase (LDH) >2 times over the upper limit of normal value; (vi) intrapulmonary lesions significantly progressed by >50% within 2–3 days; (vii) metabolic alkalosis; (viii) high sequential organ failure assessment scores; and (ix) D-dimer levels >1 mg/L at admission.

4. Differential diagnosis

In the differential diagnosis of COVID-19, we need to clarify several issues during winter and spring seasons.

First, diagnosis of pneumonia, in general, must have imaging changes. However, the WHO defines novel coronavirus pneumonia as "COVID-19" mainly because it includes mild infections, like influenza. Similar to coronavirus, we find that that many patients with influenza will not develop pneumonia, just a mild infection with upper respiratory tract or systemic symptoms. Therefore, in the sixth edition of the diagnosis and treatment plan, we proposed a differential diagnosis according to the mild infection and novel coronavirus pneumonia.

Second, mild COVID-19 should be distinguished from the common cold and influenza. The common cold primarily presents with low fever and catarrhal symptoms without seasonality. Although the entire population is vulnerable, it is typically self-limiting. Influenza can cause more systemic symptoms, such as headache and myalgia, with the most prevalent period from late November to the end of the following February. We should not ignore mild infection of SARS-CoV-2 by only focusing on other common upper respiratory tract infection viruses.

Third, for moderate, severe, and critical cases with different levels of pulmonary infiltration, epidemiological and medical history, laboratory examination, and imaging findings should be incorporated to distinguish from other types of pneumonia caused by viral (such as influenza virus, parainfluenza virus, adenovirus, respiratory syncytial virus, rhinovirus, metapneumonvirus, other coronaviruses, and other known viruses) or atypical (such as Mycoplasma pneumoniae and Chlamydia pneumoniae pneumonia) pathogens. Before the breakout of novel SARS-CoV-2, during the annual winter and spring seasons, those viruses causing pneumonia were very common, and a main cause of mortality. Therefore, especially out of Wuhan and Hubei, most cases of COVID-19 are mostly imported. The confirmation of pneumonia caused by SARS-CoV-2 should be particularly cautious. For example, after the announcement of a top-level alert in Shanghai, we received dozens and hundreds of fever cases in clinics at different hospitals. However, there were only a few COVID-19 cases in Shanghai. Most febrile patients actually had a common upper respiratory tract infection such as influenza or adenovirus pneumonia. Therefore, we have to make a comprehensive differentiating diagnosis, especially in places outside Hubei and Wuhan, and recognize the possibility of noninfectious diseases such as vasculitis, dermatomyositis, and organic pneumonia.

Fourth, special emphasis should be placed on the detection of suspected respiratory tract pathogens by means of rapid antigen detection and multiplex PCR nucleic acid detection.

Due to the lack of specificity in clinical manifestations, it is difficult to identify influenza, COVID-19, or other pneumonia only by clinical manifestations. In Fig. 4.1, the left panel shows influenza viral pneumonia and the right panel shows RSV viral pneumonia. It is difficult to identify pathogens just from CT imaging. Although the imaging features of different viral pneumonia have their own characteristics, the specificity is low. We have to establish a set of virological detection methods for clinical needs, which help us better guide clinical diagnosis and identification as well as improve the efficacy of treatment.

Fifth, clear diagnosis, isolation of patients in time, infection prevention, and avoiding excessive consumption of medical resources are vital.

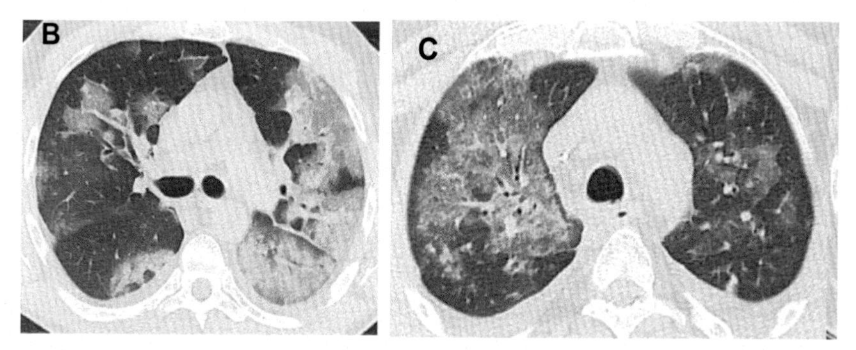

Fig. 4.1 Imaging findings of influenza virus pneumonia (left) and RSV virus pneumonia (right).

5. Cases reporting procedure

Since COVID-19 has been classified as a secondary-level infectious disease, but management is as primary level, medical workers at all levels of medical institutions should start an isolation process as soon as suspected cases are defined. After consultation with the designated senior experts, suspected cases have to be reported to the network within 2h. A SARS-CoV-2 nucleic acid test for respiratory tract or blood specimens should be conducted and the patients should be transferred to the designated hospital at the same time. Those intimate contactors who had an epidemiological correlation with the suspected case must also receive an etiological test, even if the common respiratory pathogen tests were positive.

While waiting for the response of the nucleic acid test, the patient should be transferred to a designated hospital by ambulance. It should be emphasized that the patient's condition and severity should be judged and safety is the priority. Moreover, since it is flu season, a SARS-CoV-2 nucleic acid test should also be carried out with a history of COVID-19 contact, even if the influenza test is positive. Evidence suggests that about 10% of cases suffer mixed infections at the same time. Be aware of those patients with positive influenza tests, because they may infect family on returning home, or lead to a hospital outbreak when admitted to a general ward.

6. Exclusion of cases

Two consecutive negative respiratory tract nucleic acid detection results (sampling interval of at least 1 day) can exclude the diagnosis of COVID-19. However, different points of view suggest that the period of two detections of nucleic acid at the onset of the disease could be less than 6 days, and all the nucleic acid results could be false negative. Therefore, during the onset of illness, it is recommended to have a CT scan 3 days before admission. Patients with typical imaging changes should not be allowed to return home even if two nucleic acid tests are negative.

CHAPTER 5

Treatment of COVID-19

Contents

1. Site of care

Suspected and confirmed cases should be treated in a designated hospital with effective isolation and protective conditions. The isolation condition of suspected cases should be the highest, and treatment should be carried out in a single room instead of mixed accommodation. Only confirmed cases should be admitted to the same ward, and critically ill patients should be admitted to ICU as soon as possible. At this stage, asymptomatic infected persons should also be isolated for observation. If a severe epidemic occurs in the area and medical resources are limited, mild cases and asymptomatic infected persons can be treated and observed at home, but registration and management should be carried out by the local disease prevention and control institutions and community health service centers, so as to guide, observe, and treat the quarantine at home. Moreover, the referral and

COVID-19
https://doi.org/10.1016/B978-0-12-824003-8.00005-X

transfer of severe patients should be safe, evaluated well, and no problems should be caused on the way.

2. General care

Rest in bed, strengthen supportive treatment, adequate caloric intake, pay attention to water and electrolyte balance, and maintain internal environment stability. At the same time, close follow-up should be conducted to observe patients' respiration, monitor blood oxygen saturation, and observe changes in body temperature. Attention should be paid to those patients who have changed from mild to severe or critical, especially the elderly, obese, patients with diabetes, hypertension, coronary heart disease, COPD, etc. According to the conditions, blood routine, urine routine, c-reactive protein, biochemical indicators (liver enzyme, cardiac enzyme, renal function, etc.), and coagulation function should be monitored when necessary, as well as arterial blood gas analysis and chest imaging being performed if required.

3. Antivirals

What is fundamental to an infectious disease? The bottom line is to treat the cause of disease. For COVID-19, antiviral therapy is a top priority, in both mild and severe cases. Without treatment and intervention for its cause, other treatments are very passive, so everyone is trying their best to suppress the virus at work. In fact, what we have observed is that the existence and detoxification times of coronavirus in the body are very long, and a number of body autopsies show a large number of virus particles and virus inclusion bodies in the alveolar cavity and pulmonary septum of the patient. Can other treatments work without clearing away the virus in severe patients? In addition, if the inflammation is due to excessive inflammation caused by the virus, if you do not get rid of the virus, can only immunosuppressive agents be effective? Respiratory support is an excellent tool to provide patients with valuable recovery time. But not every patient is so lucky. Lung damage in some patients is so severe that even ECMO cannot reverse it. This should remind us that etiology is the most important part of the treatment.

At present, there is no effective antiviral drug. α-interferon inhalation can be tried (adults 5 million IU each time, adding 2 mL of sterilized water for injection, twice a day) as a treatment course of at least 5 days, but no immediate effect of α-interferon atomization has been observed so far.

Notably, the interferon atomization must be performed with an air compression pump, because it can make the particles smaller—such as PM2.5 or PM5 levels—so that they can reach the alveoli. Lopinavir (200 mg) and ritonavir (50 mg) were originally used for the combination treatment of HIV, being HIV protease inhibitors and active peptide inhibitors, respectively, and the latter can increase the drug concentration of the former. Because lopinavir/ritonavir is a protease inhibitor with the ability to inhibit the formation of viral nucleic acids and has been used to treat HIV, it is also considered as a candidate drug for treating COVID-19, and has been published. This study is the first clinical trial for the treatment of COVID-19 published in the world's top medical journal since the start of the outbreak. It is also one of the few clinical trials of a drug published during the outbreak of a new infectious disease, including SARS, in the last 20 years. The accompanying NEJM editorial hailed Chinese researchers as heroes for conducting rigorous clinical trials in the midst of such a difficult outbreak. After a rigorous and scientific RCT study, it has been proved that this drug has a certain effect, but its effect is not significant. In terms of safety evaluation, the lopinavir/ritonavir group had a higher incidence of gastrointestinal adverse events. According to the results of this study, more than 40% of patients who received lopinavir/ritonavir did not have their viral nucleic acid turn negative after the 2 weeks of observation, suggesting that the antiviral effect of lopinavir/ritonavir may be limited or the course of the treatment is insufficient. On the other hand, the virus may be infectious for a long time. Some clinical observations have shown that the average time of virus turning negative is 20 days—not as short as we expected. Moreover, we initially observed that the nucleic acid turns negative after more than 30 days. In the future, detailed observations should be made, including multisite sampling, to see if viral nucleic acids persist for a longer time. This is the disease pattern that we need to understand further. Does the course of lopinavir/ritonavir treatment need to be extended? An extension may be necessary until viral nucleic acids become negative. According to data from the clinical study, lopinavir/ritonavir is effective, has some side effects, but can be tolerated; in addition, the course of the treatment may need to be longer.

Avoiding adverse reactions is quite important for the use of lopinavir/ritonavir; these include liver function damage, diarrhea, and nausea in patients with obvious gastrointestinal symptoms. It is also necessary to avoid interactions with other clinical drugs, especially to avoid combined use of simvastatin and erythromycin, which may be used by some elderly patients complicated with chronic obstructive pulmonary disease, hypertension, hyperlipidemia, and other diseases. It is necessary to pay attention to the interaction between drugs so as not to aggravate liver function damage as combined utilities. When lopinavir/ritonavir is used in combination with erythromycin, it may cause some symptoms of arrhythmias with prolonged QT intervals.

Until the results of a rigidly designed, randomized, double-blind trial are known we cannot predict how effective or ineffective remdesivir will be, or how much it can suppress the COVID-19 virus. Whether remdesivir is ultimately shown to be effective or not will have direct implications for clinical treatment.

4. Glucocorticoids

At present, there is great controversy about the application of hormones: some think that hormones are effective, while others think that hormones do more harm than good. Although there is controversy, it is undeniable that hormones are commonly used in the treatment of viral pneumonia. Evidence-based medicine for the treatment of viral pneumonia is controversial. Most of these studies are retrospective studies with poor comparability and insufficient evidence levels. As we all know, the level of evidence in retrospective studies is inadequate and its comparability is poor. A good study requires large-scale, multicenter, randomized, double-blind, controlled studies, but it is too difficult to do according to this design. Hormone dosage, timing, usage, and duration vary. For example, patients have increased mortality after using hormones, mainly because many patients are already very sick when using hormones, while mildly ill patients do not need to use hormones. This leads to an unscientific conclusion that the mortality of patients who use hormones is high. The high mortality rate is not caused by hormone treatment, but it may be caused by severe illness. At the present stage in the fever clinic and ward, because some patients cannot be admitted to the hospital in time, or they come to see the doctor when their condition is very serious, these patients may not be able to get timely treatment, and may die within 2 or 3 days. At this time, although doctors use hormones, the effect has not been revealed. If we do research at this time, we will find that the mortality rate of these patients increased after the use of hormones. So, at this time, our conclusion may not be scientific. In addition, if the usage and dosage of hormones are not standardized, the conclusions of the study may also be biased. For patients with novel coronavirus pneumonia, how should hormones be used? From our experience, hormones should not be used in patients with early, mild symptoms, such as within 7 days of infection, or between 7 and 10 days, as this will accelerate the replication of the virus. When the patients are at the peak of the disease—that is, if the patients have sustained high fever, obvious dyspnea, hypoxemia, and obvious progression of chest imaging lesions after 7–10 days—it may be the best time to use steroids. In some patients with COVID-19, the imaging manifestations are of organizing pneumonia, which is sensitive to hormones, and this is a very good time to use steroids. As described below, there are two cases of hormone treatment to help patients survive from the period of hypoxemia.

Case 1: 77-year-old, male, admitted to hospital on January 10, 2020. He became ill in January. He was admitted to hospital after 1 week of fatigue and 4 days of fever. Past history: diabetes for 22 years, treated by long-term oral metformin; hypertension history for 20 years. Fig. 5.1 shows his chest image on January 9, 2020, with multiple ground-glass shadows in both lungs.

By January 14, 2020, the ninth day of disease onset, the patient's chest CT showed a significant progression in ground-glass shadows (Fig. 5.2). The study found that many cases tend to progress during 7–10 days from onset of the disease, or even after 10 days, but the degree of progression varies. Some patients presented with increased dyspnea and respiratory failure, while others presented with increased imaging findings, but no obvious respiratory failure. The patient's chest CT scan on January 9 and January 14 showed significant exacerbation within 5 days, clinically manifested with

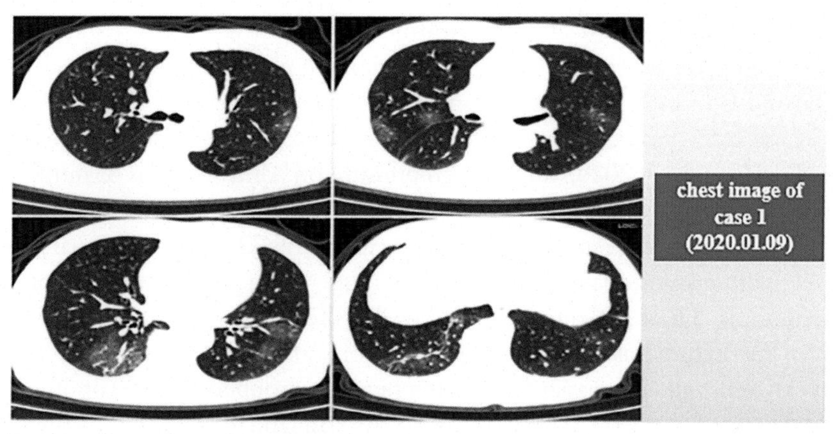

Fig. 5.1 Chest image of case 1 (January 9, 2020).

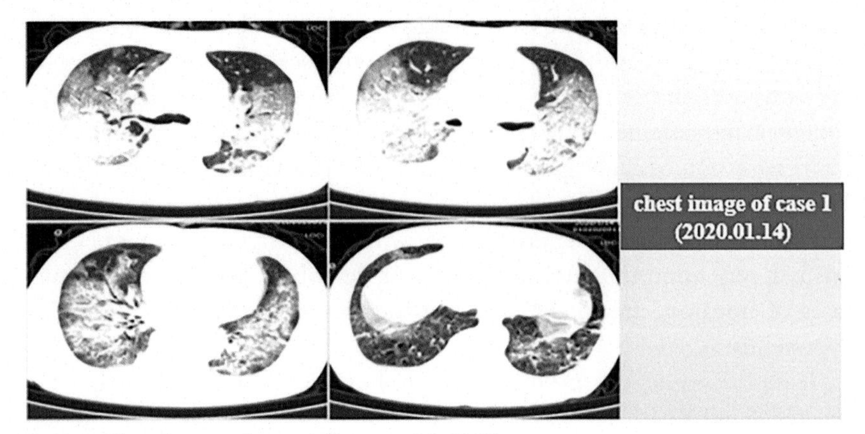

Fig. 5.2 Chest image of case 1 (January 14, 2020).

fever, dyspnea, and hypoxemia. So for this patient, we applied the ventilator in time.

As shown in Fig. 5.3, from January 10, 2020, we applied methylprednisolone with an initial dose of 60 mg per day, but the condition did not improve. So we increased the hormone level to 80 mg per day (40 mg per time, twice a day), but the patient's condition continued to deteriorate. On January 14, the patient developed respiratory failure. We applied methylprednisolone to 160 mg per day (80 mg per time, twice a day), and the patient's condition gradually controlled and remained stable. Then we gradually reduced the dose of hormone.

Fig. 5.4 is the chest CT on January 18, 2020. After 4 days of treatment, including increased hormone doses, use of a noninvasive ventilator, and other antiinfective and antiviral treatments, the inflammatory exudation began to decrease. After overcoming this difficult period, the patient recovered and was discharged from hospital on January 29.

Case 2: 77-year-old, male, with fever for 10 days, and admitted to hospital on January 7, 2020. The oxygen pressure on admission was 65 mmHg and the patient was immediately transferred to ICU and treated with noninvasive mechanical ventilation. Fig. 5.5 is the chest CT image on January 7, 2020, with ground-glass shadows in both lungs, characterized as organizing pneumonia. These patients responded well to the hormone. The patient was given methylprednisolone 80 mg per day for 7 days, and the dyspnea was quickly relieved. The patient's body temperature dropped and the condition was stabilized. A chest CT scan was performed on January 19, showing obvious absorption of the lesion (Fig. 5.6).

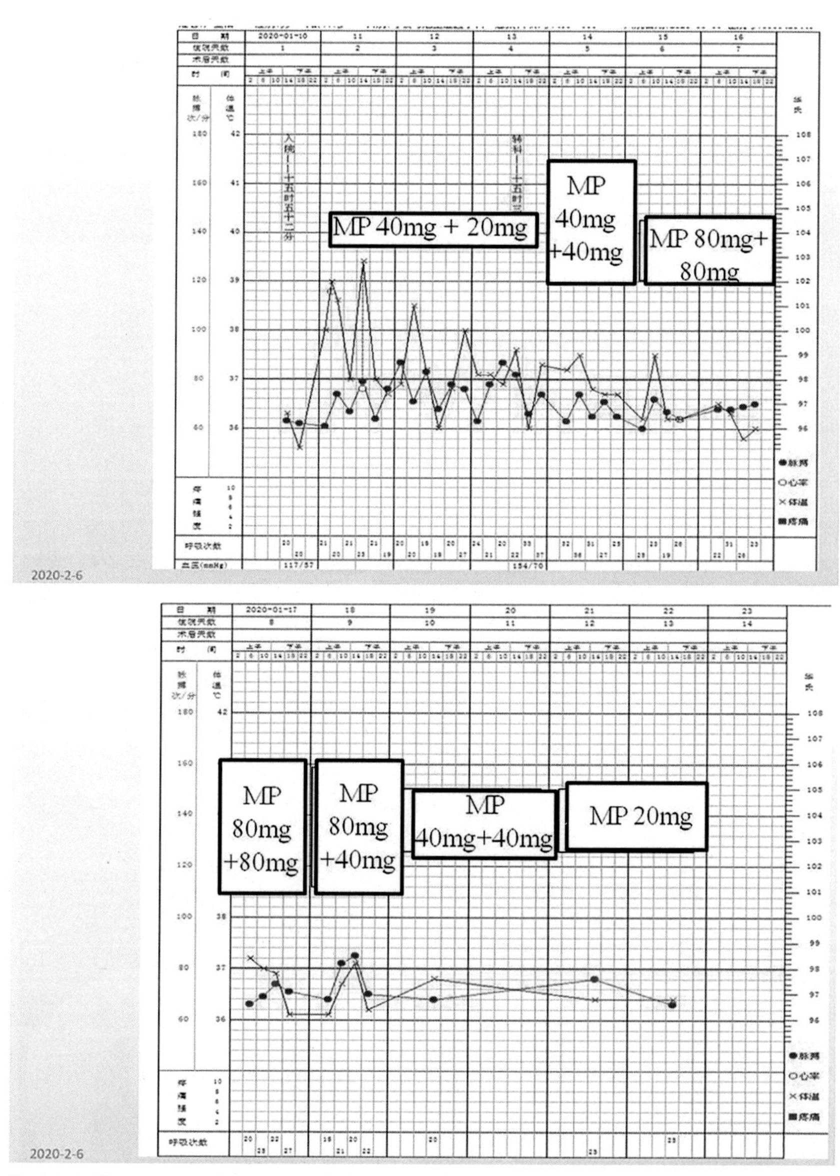

Fig. 5.3 Hormone utilization and changes of vital signs in case 1 (January 10, 2020). MP: methylprednisolone.

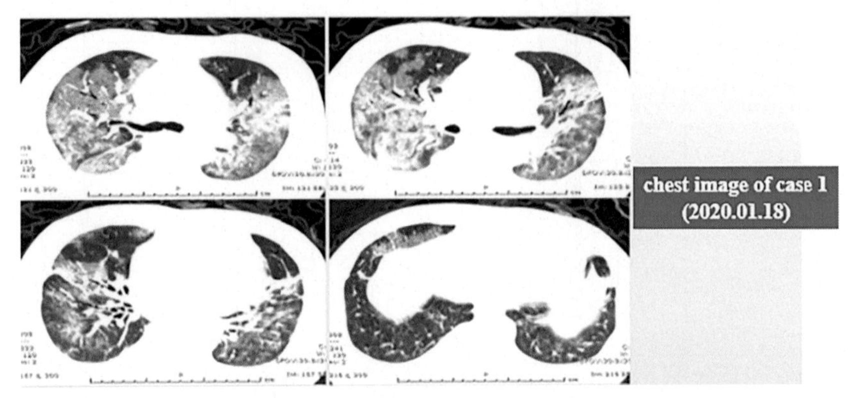

Fig. 5.4 Chest CT image of case 1 (January 18, 2020).

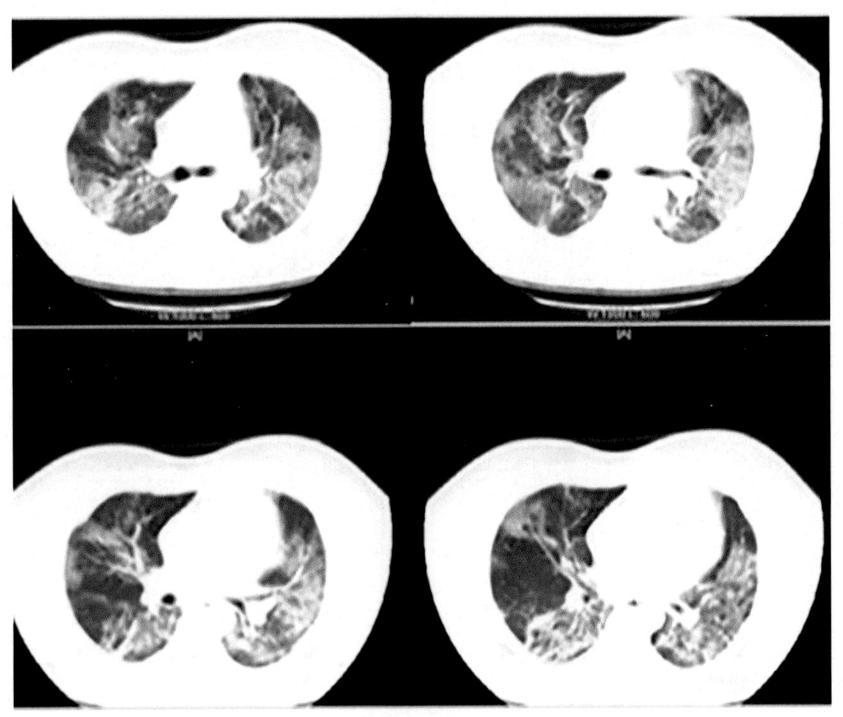

Fig. 5.5 Chest CT image of case 2 (January 7,2020).

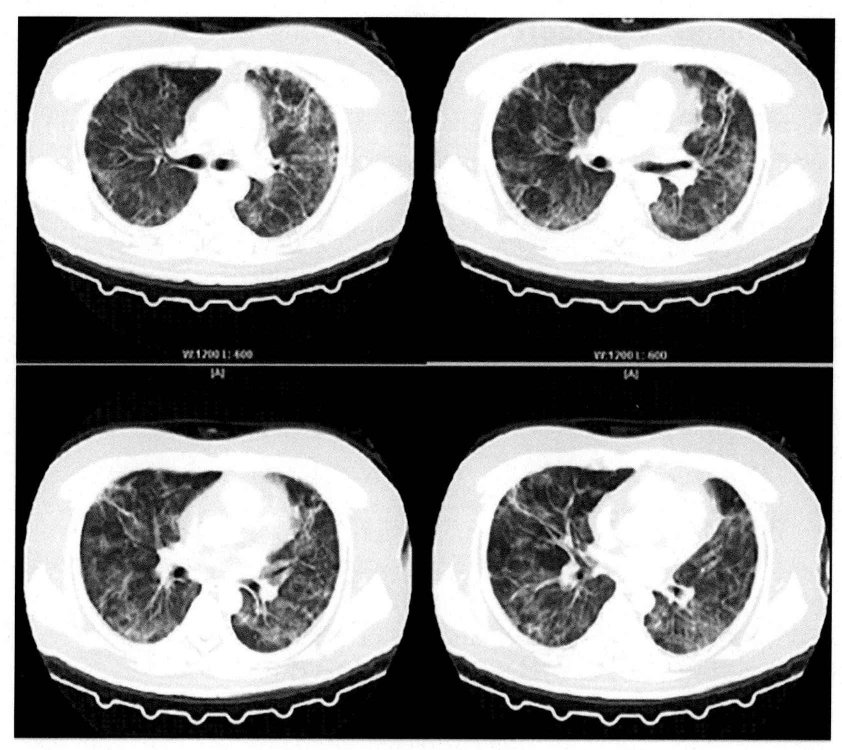

Fig. 5.6 Chest CT image of case 2 (January 19, 2020).

Hormone use, if appropriate, can be beneficial for patients. Hormones will inhibit the immune function. Therefore, in the early stages of the disease, immunosuppressive agents should not be used. After 7–10 days, hormones can be added in cases with persistent fever, obvious dyspnea, hypoxemia, progressive imaging progression, or SOPs-like changes. Before using hormones, doctors must evaluate the cellular and humoral immune function of patients, and must evaluate the viral load, the patient's lymphocyte, T cell subsets, and NK cells. After comprehensive assessment, we can use hormones. If the patient's cellular immune system is compromised and the virus is still replicating, glucocorticoids should be used with care. The changes of lymphocyte, c-reactive protein, oxygenation index, and imaging manifestations should be closely observed after glucocorticoid administration. Generally, the period of glucocorticoids treatment should be 3–5 days, then the dosage should be reduced slowly after the improvement of illness, and the administration can be stopped. Large doses and prolonged time of usage should be avoided.

5. Respiratory support

5.1 General principles

According to the change of arterial oxygen saturation or peripheral capillary oxygen saturation, effective oxygen therapy should be given in time, including oxygen delivery by nasal catheter or face mask, and high-flow nasal cannula (HFNC) therapy, with noninvasive or invasive mechanical ventilation if necessary.

The respiratory support methods for severe COVID-19 patients are basically high-flow nasal cannula (HFNC) oxygen therapy, noninvasive mechanical ventilation, and invasive mechanical ventilation at the early stage, and extracorporeal membrane oxygenation (ECMO) at the advanced stage.

5.2 HFNC

High-flow nasal cannula is an effective way to correct patient's hypoxia. Patients with mild ARDS (oxygenation index 200–300 mmHg) or mild COVID-19 are recommended for use of HFNC, but its efficiency in improving hypoxia is inferior to that of a bilevel positive airway pressure (BiPAP) ventilator. Making a choice between HFNC and noninvasive ventilation depends mainly on patients' tolerance and response to oxygen therapy: HFNC should be considered when patients are intolerant of a BiPAP ventilator or show rapid improvement in hypoxia once receiving HFNC therapy. Noninvasive ventilation is preferred when applicable and available.

Initial setting for HFNC: FiO_2 is set to 100%, with a slow increase in flow rate from 30 L/min to 50 L/min. Consider using the ROX index [$SpO_2/(FiO_2*RR)$] to predict the success of HFNC.

(1) Evaluate once every 2 h. An ROX index higher than 3.85 predicts a low risk of endotracheal intubation and high probability of success of HFNC, and therefore HFNC treatment is recommended to be continued.

(2) If the ROX index at 2 h is less than 2.85, immediate endotracheal intubation should be considered.

(3) If the ROX index at 2 h ranges between 2.85 and 3.85, continue HFNC treatment and reevaluate the ROX index at 6 h: continue HFNC treatment if the index at 6 h >3.85, and perform immediate endotracheal intubation if <3.85. Evaluate the Rox index again at 12 h: an index of >4.88 indicates great probability of success of HFNC, and an index of <4.88 suggests immediate endotracheal intubation.

The ROX index is used to determine optimal timing of the switch from HFNC to endotracheal intubation. It is safer to observe for the first 2 h, because longer observation time indicates possible worse effect of HFNC, which may lead to a delayed intubation. Additionally, it is necessary for the medical staff to observe at the bedside for a period of time to calculate the ROX index. Whether the endotracheal intubation timing of critical COVID-19 patients can be determined based on the ROX index only needs further observation and verification. The best threshold still needs to be explored. In general, the efficacy of HFNC on severe ARDS is extremely limited. The timing of endotracheal intubation should not be delayed due to a transient improvement in oxygenation index resulted from HFNC treatment.

5.3 Noninvasive mechanical ventilation

Attention should be paid to the use of noninvasive ventilators. Some hold that if a 2-h noninvasive ventilation does not yield an improvement in the oxygenation index, one should immediately change to invasive mechanical ventilation. Others insist that we should only consider invasive ventilation when noninvasive ventilation has proved to have no effect. Actually, appropriate use and careful adjustment could ensure effective noninvasive ventilation and save patients' lives.

Respiratory failure in COVID-19 patients has resulted from ARDS or interstitial pneumonia. Invasive mechanical ventilation may attenuate ventilation/perfusion ratio imbalance, but could not improve diffusing capacity due to interstitial involvement. Worse, oxygenation was often observed when switching noninvasive ventilation to invasive ventilation. Inability to extubate within a short period may lead to prolonged invasive mechanical ventilation, and therefore increase the risk of ventilation-associated pneumonia, which contributes to high mortality. Some studies found that the mortality rate of COVID-19 patients using invasive mechanical ventilation is more than 90%; however, the use of a noninvasive ventilator requires medical staff to stay bedside for repeated adjustment, including checking the appropriateness of parameters and absence of gas leakage. Most noninvasive ventilation failure is associated with inadequacy of adjustment and premature discontinuation after a first failed attempt.

Here is a representative case. A 70-year-old female patient with COVID-19 was admitted for hospitalization. Her blood oxygen saturation dropped below 80% and stayed around 80% for more than 2 days after receiving noninvasive ventilation. Rather than performing intubation on this patient, according to the previous point of view, our nurse stayed at the bedside to adjust carefully the parameters of the noninvasive ventilator. With the help of antiviral drugs and steroids, blood oxygen saturation improved to 90% on the fourth day postnoninvasive ventilation, and even reached 97%–98% on the seventh day. If the patient had been intubated, it's very likely that she could not have been extubated within such a short period and a series of complications might have appeared. Therefore, time and effort should be put into ensuring successful application of noninvasive ventilation.

It is worth noting that patients with shortness of breath may not be able to cooperate with the noninvasive ventilator, so it is necessary to evaluate the patient in advance. If the patient is at risk of the disease progressing, early application of noninvasive ventilation should be considered. In some cases, patients' blood oxygen saturation on nasal catheter may reach 94%–95%, which seems to be a good condition. However, if a noninvasive ventilator is applied when the patient's respiratory rate is not that fast, it's easier to achieve good tolerance and patient-ventilator synchrony, so that when the patient's condition deteriorates, the ventilator will take effect. We need to raise awareness of the importance of introducing a noninvasive ventilator early.

Regarding initial settings of noninvasive ventilation, a patient with strong spontaneous respiratory drive has high minute ventilation, and therefore a PEEP of 5–8 cmH_2O is sufficient, and it's not recommended to have the patient on a high level of PEEP at the early stage, such as 10 cmH_2O, 12 cmH_2O, or 14 cmH_2O. An air leak that occurs around the mask will aggravate the patient's condition due to insufficient oxygen supply, and should be attached with utmost care. Another critical parameter is oxygen concentration. The recommended initial FiO_2 for noninvasive ventilation, if applicable, is 1.0, with a gradual decrease, rather than an increase starting from 0.5 or 0.6, because patients' early demand for oxygen supply is great.

There are many predictors for noninvasive ventilation failure, and special attention should be paid to changes in states of consciousness and hemodynamic abnormalities: loss of consciousness or occurrence of irritability after noninvasive ventilation indicates the urgent need for endotracheal intubation; a significant drop in blood pressure or hemodynamic instability when receiving noninvasive ventilation also requires timely intubation. After receiving noninvasive ventilation, some patients still maintain consciousness, and present attenuated dyspnea and increased oxygenation index, with normal hemodynamic parameters. However, severe respiratory muscle fatigue develops in those patients, and weaning from noninvasive ventilation for a short period of time to eat and drink may induce a rapid decrease in oxygen saturation, which is slow to recover even when noninvasive ventilation is continued. Such patients should be actively assessed for the necessity of invasive mechanical ventilation.

A series of studies on ARDS have shown that high tidal volume under noninvasive ventilation is an independent risk factor for predicting noninvasive ventilation failure, though no consensus has been reached on the threshold value, which ranges from 8 mL/kg, through 9 mL/kg, to 9.5 mL/kg of ideal body weight. Minute ventilation and tidal volume are two important factors for prediction. Minute ventilation is calculated as tidal volume (12 mL/kg of ideal body weight) times of measured respiratory rate. If the minute ventilation is maintained around 12 L, the tidal volume ranges between 8 and 10 mL/kg of ideal body weight, the respiratory rate is below 25 breathing per minute, and the oxygenation index is stable, noninvasive ventilation should continue to be used; otherwise, endotracheal intubation should be considered.

5.4 Invasive mechanical ventilation

5.4.1 The lung protective ventilation strategy should be strictly enforced

The initial setting of tidal volume is 6 mL/kg of ideal body weight, regardless of the ventilator mode, pressure-controlled or volume-controlled ventilation. We are inclined to choose pressure-controlled ventilation for COVID-19 patients, because 70%–80% of patients present excessive purulent and bloody airway secretions after endotracheal intubation. If it is not possible to clean the airway rapidly after endotracheal intubation, increased airway resistance will render it difficult to achieve an inhaling peak pressure of $<42\,cmH_2O$ or plateau pressure of $<30\,cmH_2O$ under volume-controlled ventilation.

Plateau pressure should be limited to $<30\,cmH_2O$. If the plateau pressure exceeds $30\,cmH_2O$, gradually reduce the tidal volume at a speed of 1 mL/kg of body weight, until the plateau pressure is $<30\,cmH_2O$ or the minimum of tidal volume is reduced to 4 mL/kg of body weight. In the meantime, ensure adequate alveolar ventilation and increase the respiratory rate to improve clearance of carbon dioxide (usually 5 breaths/min per change). In addition, pay attention to the ventilator waveform. Ideally, the patient's end expiratory flow will return to zero, which indicates that there is no hyperventilation or excessive gas trapping. The presence of end expiratory flow reflects incomplete expiration, which requires a reduction in respiratory rate or an adjustment of the inspiration expiration ratio to extend expiration time.

5.4.2 FiO_2 is initially set at 100%

Initial PEEP settings are suggested in the ARDSnet FiO_2-PEEP table. PEEP corresponding to a FiO_2 of 6 L/min or 0.7 L/min is chosen as the initial setting, which should be $20 \times FiO_2 \pm 2$. In some cases, especially those having delayed intubation, lung injury will aggravate if the initial PEEP setting is too high, and therefore it is generally at a level of 10–12 cmH_2O. It is not necessary to achieve a blood oxygen saturation (SpO_2) at 100%, instead, SpO_2 maintained at 88%–95% with early partial pressure of oxygen no less than 60 mmHg is sufficient for patients' needs. Previously, we observed that the partial pressure of oxygen in a patient who underwent emergency intubation as a rescue measure with a FiO_2 of 100% increased from 20 mmHg to 50 mmHg at the early stage, and gradually increased to 70–80 mmHg after 2 or 3 h. Patience is required for management of this kind of case.

Early use of prone ventilation does work. Many patients who had an oxygenation index of 50 mmHg showed significant improvement within 12 h and even reached an oxygenation index of 250 mmHg after endotracheal intubation with early prone ventilation. Almost all patients who were given early prone ventilation after intubation presented significantly improved in terms of their oxygenation index, but the efficacy of prone ventilation gradually declines after 96 h. In our department, prone ventilation is required to be initiated 3 h postendotracheal intubation when oxygen supply is stable, and should last at least 18 h a day.

After endotracheal intubation for invasive ventilation, pay attention to conservative fluid management and initiate prone ventilation when required.

Ensure adequate sedation and analgesia at an early stage. Considering the insufficient oxygen supply for ARDS lungs, achieving a deep sedation state helps to reduce the patient's whole body oxygen consumption. Additionally, early adequate sedation and analgesia can reduce the inflammatory storm brought on by excessive spontaneous breathing and avoid ventilator-induced lung injury.

Other aspects mainly include blood glucose management, blood pressure management, nutrition support, and so on.

5.5 ECMO

The principle of initiating ECMO is that the initiating timing should be based on static compliance (Cstat) and dynamic compliance (Cdyn). If possible, measure esophageal pressure and transpulmonary pressure.

The timing of ECMO initiation currently set by us is not necessarily accurate and is only for reference. For patients undergoing prone ventilation for 24h, if the oxygenation index has no improvement and stays below 100mmHg, and CO_2 retention occurs with a $PaCO_2$ of >50mmHg, initiating ECMO will be considered. In addition, if the patient develops an early and persistent intractable metabolic acidosis with pH < 7.2, ECMO initiation is also suggested. A large body of literature suggests that for patients receiving mechanical ventilation for more than 7 days with high levels of oxygen delivery, ECMO is of little significance and therefore is not recommended.

Once connected to the ECMO, the ventilator settings must be adjusted: respiratory rate is suggested to be at 8–10 breaths/min and no more than 20 breaths/min; adjust PEEP based on patient-specific changes in condition; if possible, reduce FiO_2 to <0.4 to avoid oxygen toxicity. Pay attention to management of all kinds of catheters to avoid bloodstream infection, especially for immunosuppressed patients.

5.6 Respiratory support and respiratory therapists in the treatment of severely and critically ill patients

The focus of respiratory support includes ordinary oxygen therapy (administration of oxygen via nasal cannula or face mask), delivering high flow oxygen through a nasal cannula, noninvasive ventilation, invasive ventilation, and ECMO. Airway management should also be given significant importance, including sputum suction and management of artificial airways. Currently, respiratory therapists (RTs) are urgently needed in critical care areas to help doctors and nurses to carry out more specialized respiratory support.

In the treatment of novel coronavirus pneumonia, RTs should play an important role in providing respiratory treatment, and pay special attention to ensure each treatment is standardized. Currently, there is still much room for improvement in oxygen supplement methods, including usage of nasal catheter, HFNC, and especially noninvasive ventilation. The efficacy of noninvasive ventilation depends largely on the users' depth of knowledge and skillfulness in device settings. Clinical judgment should be adapted to the ability and level of your team in providing respiratory support, and you should identify the optimal timing for treatment.

In terms of airway management, COVID-19 patients are very likely to present airway mucus obstruction, with pathological changes of peripheral small-airway sputum bolt. Wetting the airway with water, which helps to remove phlegm, is of critical importance. In terms of application of apophlegmatic drugs, combining intravenous and oral medications is suggested, such as N-acecysteine and anningpai (eucalyptol, limonene, pinene) enteric soft capsules for oral route, and ambroxol for intravenous route.

6. Convalescent plasma therapy

In addition to drugs, another potential antiviral treatment is convalescent plasma therapy, a relatively old therapy. Convalescent plasma is an important therapeutic option for humans to deal with emerging infectious diseases, especially when there is limited research on novel drug development. Convalescent plasma should theoretically be effective, as plasma derived from recently recovered patients contains certain neutralizing antibodies, which may be able to fight, neutralize, and eliminate the virus. The application of plasma therapy in emerging infectious diseases, which currently include SARS, avian influenza, MERS, and others, has a history of more than 100 years. Some clinical trials have suggested that convalescent plasma therapy is effective, while no obvious effect was observed in studies on Ebola virus disease. No specific treatment for COVID-19 is currently available, and therefore convalescent plasma treatment could be an important option for rescuing severe patients.

Compared with the convalescent plasma therapy of the 19th century, that of the 21st century has contemporary methods of observation and treatment that have earned it more attention at the level of plasma antibodies, such as receptor binding domain antibodies and neutralizing antibodies. Convalescent plasma preparation requires strict procedures to ensure the safety of plasma and adequate titer of plasma neutralizing antibodies. Some biopharmaceutical enterprises that are qualified to prepare plasma have actively participated in developing convalescent plasma for clinical treatment. Ruijin Hospital assisted Wuhan frontline hospitals to carry out a plasma treatment study in 10 severe COVID-19 cases. After convalescent plasma transfusion, all patients' clinical symptoms improved within 24–48 h, and significant improvements in inflammatory indicators and pulmonary radiological images were observed, especially the obvious absorption of ground-glass opacity.

However, this study is exploratory and moderate patients as well as severe patients getting better do not necessarily need plasma therapy. Plasma therapy is recommended to be used in severe patients with progression, especially those without ARDS. The occurrence of ARDS indicates the patient has been severely ill and transfusion of convalescent plasma with high titer of neutralizing antibodies may still not work. The suggested optimal timing for convalescent plasma transfusion is early after the patient progresses to the severe or critical stage. Transfusion before the occurrence of ARDS helps to prevent progression to critical illness. Meanwhile, strictly prepared plasma is required by Chinese laws and regulations on blood transfusion. Relevant biopharmaceutical enterprises must comply fully with China's plasma preparation technical standards to ensure the quality of plasma and patients' safety. Another controlled study of convalescent plasma treatment is currently recruiting patients in Wuhan. It is hoped that 50 patients will be enrolled for convalescent plasma therapy and 50 patients for SARS-CoV-2 nonimmune plasma therapy. This study aims to obtain further medical evidence of plasma therapy, and provide an objective and comprehensive evaluation on its efficacy among COVID-19 patients, especially those in early, severe, or critical conditions.

Recently, the National Health Commission published the "Guidelines on Convalescent Plasma Therapy in COVID-19 Patients (Trial First Edition)," providing eligibility requirements and criteria for both recipients and donors, with the aim of ensuring the appropriate and standard application of plasma therapy.

7. Cytokine-targeted therapy

Considering the cytokine storm in COVID-19 patients, a clinical study on talizumab—a monoclonal antibody that specifically binds to the interleukin-6 receptor—was conducted with the joint efforts of the Chinese University of Science and Technology, School of Life Sciences and Medicine and its affiliated hospitals. In the published data of the current phase 1 clinical trial, the body temperature of 11/14 COVID-19 patients returned to normal within 24 ho with improvement in the oxygenation index; four patients presented absorption of lesions in chest CT images, and one critical case has been successfully weaned off from endotracheal intubation. The follow-up data of 14 patients will be updated in the future.

8. Venous thromboembolism prophylaxis and treatment

Hypercoagulation and hyperfibrinolysis are common in COVID-19 patients, as evidenced by clinical presentations and pathological findings of thrombosis. No contraindications means a certain degree of indication of anticoagulant therapy. Treatment regimens could be made based on a patient's disease severity. A patient with a confirmed diagnosis of VTE could be given a treatment dosage, and those without VTE could be given a prophylactic dosage, and adjustments made according to age and other hemorrhagic risks. It is necessary to monitor the risk of heparin-induced thrombocytopenia (HIT) when administering heparin.

9. Antibacterial treatment

Principally, patients with mild COVID-19 do not need antibacterial therapy, or any special treatment. They are mostly self-limiting within 2 weeks after symptomatic treatment. It is estimated that more than 90% of patients with mild forms of the disease will improve or become self-healing, and therefore they do not require antibacterial drugs.

The progression of COVID-19 is relatively slow, unlike influenza, which may progress to severe pneumonia requiring hospitalization 3–4 days after onset of illness. The average intubation time of severe COVID-19 cases is 11 days. Secondary bacterial infection is rare among patients receiving invasive ventilation for 2 or 3 days, but it might occur in patients under ventilation for more than 3 days, such as carbapenem-resistant *Enterobacteriaceae* (CRE, such as *Klebsiella pneumoniae*), *Acinetobacter baumannii*, and *Staphylococcus aureus*. In general, patients admitted for a short period of time are not likely to have bacterial infection. It is recommended to avoid unnecessary or inappropriate use of antimicrobial drugs at the early stage of hospitalization, especially combination use of broad-spectrum antibiotics.

Novel coronavirus pneumonia is also a community-acquired pneumonia, and the recommendations on antimicrobial treatment in "Guidelines on Diagnosis and Treatment of Community-Acquired Pneumonia in Adults," issued in 2016, should be followed. For patients with confirmed diagnosis of novel coronavirus pneumonia and suspected infections of other pathogens, it is not inappropriate to empirically use antimicrobial drugs. Therefore, from a practical point of view, try not to use antimicrobial drugs for patients with mild cases without obvious evidence of bacterial infection. When bacterial infection cannot be excluded, it is recommended to decide empiric treatment according to local epidemiological resistance profiles and avoid using broad-spectrum antimicrobial drugs. For the treatment of patients with severe forms of the disease, empirically use antibiotics to ensure broad-spectrum of coverage after rapid collection of diverse specimens, followed by discontinuing or deescalating antibiotics based on microbiological culture results and switching to narrower-spectrum antibiotics to target the identified pathogens.

Secondary bacterial infection tends to occur after viral infection, so bacterial monitoring should be strengthened during COVID-19 treatment. If there is any evidence of secondary bacterial infection, such as expectoration of purulent or yellow sputum, or elevated procalcitonin, prescribe antimicrobial drugs and carry out personalized treatment according to the type of infection, comorbidities, and liver and kidney function; try not to use broad-spectrum antimicrobial drugs. Inappropriate and overextended use of broad-spectrum antibiotics might lead to secondary fungal infection. When suspecting secondary fungal infection, pathogen monitoring including G test and GM test should be considered. For patients with lymphopenia, such as those with lymphocyte count below 300–400/uL, it should be considered whether to take measures for prevention of *Pneumocystis Carinii* pneumonia. When dealing with infections, remember to take all important issues into account, including etiological aspects.

10. Standards for discontinuation of isolation and hospital discharge

The following four criteria need to be met for a patient to be eligible for hospital discharge or discontinuation of isolation:

① Body temperature returns to normal for more than 3 days.

② An obvious improvement in respiratory symptoms is observed.

③ Chest CT imaging shows a marked improvement in acute exudative lesions.

④ Two consecutive negative results of nucleic acid test are obtained in respiratory specimens (the sampling interval is greater than 1 day).

Patients are allowed to go home and return to work or school according to the fifth version of the "Guidelines on Diagnosis and Treatment of Covid-19," which was issued by the National Health Commission. However, according to the sixth version, patients who have met the standards of discharge may still not be able to go home as required. For example, patients who have been discharged from the hospital and returned to their community or place of residence must be further quarantined for another 2 weeks. This has two main purposes: first, to restore their own resistance, and second, to avoid transmitting the virus to others in cases where patients are still contagious within this 2 weeks.

CHAPTER 6

Prevention and disease control of COVID-19

Contents

1. Principles

The control of infectious disease is more dependent on prevention than on treatment.

The first task is to isolate the source of infection. Suspected patients, mildly affected patients, and close contacts of confirmed cases should be placed under medical observation. No matter whether there is an etiological diagnosis or not, suspected patients should be kept in strict isolation. It is difficult to identify the source of infection completely unless compulsory measures are taken, such as door-to-door screening. Therefore, the focus of

COVID-19 https://doi.org/10.1016/B978-0-12-824003-8.00006-1

prevention is how to cut off the transmission routes. Given that droplet transmission and contact transmission appear to be the main routes of transmission of COVID-19, the general public need to refrain from going outdoors as much as possible, wear masks in public, and keep good hygiene including frequent handwashing, and wiping and disinfecting door handles and elevator buttons. It is recommended to stop using central air-conditioning because COVID-19 may also spread through aerosol transmission.

2. Use of personal protective equipment

2.1 Selection of PPE

Medical personnel should take adequate measures to prevent nosocomial infection, which not only protects them, but also better protects patients. On January 26, the National Health Commission of the People's Republic of China issued the Guide of Common Medical Protective Equipment in Novel Coronavirus Infected Pneumonia (Trial).

While coping with infectious diseases transmitted by droplets, use masks at the medical protection level. First of all, the waterproof coating of the mask is key: the common antismog and antidust face masks perform relatively poorly for waterproof functions. In addition, pay attention to the filtering capacity and sealing performance of the mask. Based on the abovementioned factors, patients should use medical surgical masks; medical staff must use medical protective masks when they enter quarantine areas.

The impermeability of protective clothing should also be emphasized and double-layer protection used when necessary. The latest research suggests that, besides droplet and contact transmission, it is possible for COVID-19 to spread through the fecal-oral route, and therefore the protection requirement is considered to be higher. Because there may be aerosol transmission, it is necessary to have a full three-dimensional and omnidirectional protection in high-risk areas, including goggles and face masks.

Medical staff should evaluate the risk level according to their positions, and then use a corresponding level of protection. Higher risk operations require higher levels of protective equipment. High-risk operations contain operations that involve contact with patients' blood, body fluids, secretions, etc., or operations that may generate aerosols, such as performing tracheal intubation and bronchoscopy, collecting pharyngeal swabs, suctioning sputum, and giving oral nursing. Medium-risk operations refer to direct contact with the patient without performing the abovementioned high-risk operations, and include physical examination, injection, puncture, etc.

While conducting medium-risk operations, the protection of medical personnel must include work clothes, gowns, caps, goggles, and medical surgery masks or medical protective masks. Low-risk operations consist of operations that only involve indirect contact with patients and enable personnel to keep a certain distance from the patient, such as prescription of drugs to patients and interviewing patients (where no physical examination is required).

2.2 Use of masks

Whether it is a medical surgical mask or a medical protective mask, it is particularly important to wear it correctly. Fig. 6.1 shows how to wear a medical surgical mask. Medical surgical masks can block particles larger than 5 μm, so can curb the spread of droplets. The medical surgical mask is disposable, and its effective protection time is 4 h, which can be extended slightly. The mask must be replaced immediately once contaminated.

Fig. 6.2 shows the method of wearing a medical protective mask. In addition to the five steps listed above, there is an essential sixth step: perform a marginal tightness check after wearing the mask as shown in the figure. Although medical protective masks can block particles as small as 0.3 μm, if we just put on the mask but do not wear it well, particles of 3, 10, or even 30 μm will go in; then it is equivalent of wearing no mask.

Some medical personnel wear a surgical mask inside a medical protective mask. This is not permitted, as it will reduce the effectiveness of the protective mask by diminishing the tightness of the protective mask. If you intend to keep using the medical protective mask for a period of time while leaving the contaminated area, you could wear a surgical mask outside the medical

The medical surgical mask procedure is as follows:

•Cover the nose, mouth, and chin with the mask. The lower tie-wraps of the mask are tied behind the neck, and the upper tie-wraps are tied to the middle of the top of the head.

•Put your fingertips on the nose clip, press inward with your fingers from the middle position, gradually move to both sides, and shape the nose clip according to the shape of the bridge of the nose.

•Adjust the tightness of the lace.

Note:

•The nose clip should not be pinched with one hand.

•Single-time use of surgical mask (effective protect time: 4 h).

•Masks should be replaced promptly after becoming wet or contaminated with blood or body fluids.

Fig. 6.1 Wearing methods of medical surgical masks.

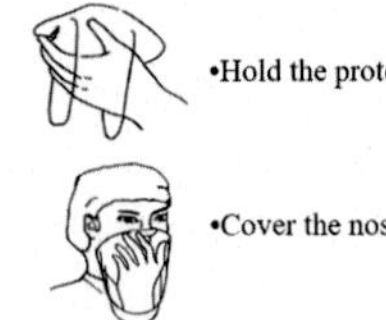

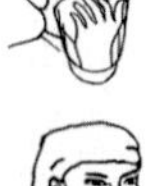

Fig. 6.2 Wearing method of medical protective mask.

protective mask. When you leave the contaminated area, remove the surgical mask and continue to use the inner protective mask. If this is not the case, there is no need to wear a surgical mask outside the medical protective mask. Hand hygiene is required after removing the mask.

2.3 Use of gloves

Common examination gloves are relatively easy to damage and difficult to fix. When we enter high-risk areas, it is better to wear two or even three layers of surgical gloves. However, wearing gloves cannot replace hand hygiene. When taking off protective equipment, hand hygiene should be performed at each step and between each layer. Do not touch unnecessary positions and items after putting on latex gloves, otherwise it will cause the spread of pollution.

2.4 Use of goggles and protective face shields/face screens

The use of goggles and protective face shields/face screens is necessary when performing operations involving splashing of body fluids, blood, etc., as these can protect the eye mucosa. These kinds of protective equipment are generally reusable as long as we ensure that the disinfection procedure meets the requirements. It is forbidden to leave the diagnosis and treatment area while wearing goggles and protective face shields. The three objects do not need to be used at the same time.

2.5 Use of gowns and protective clothing

In terms of gowns and protective clothing, appropriate protective equipment should be used according to the exposure risk of different working environments. Correct wearing technique, and especially correct undressing, is extremely important. It is critical to take PPE off in a standardized way and prevent pollution during this process. What is more, you should not leave the isolation ward while wearing medical protective clothing. In addition, shoe covers should be used in specified areas only. Shoe covers are required when entering a contaminated area from a semicontaminated area, as well as when entering a negative pressure ward from a buffer area. Conversely, shoe covers should be removed when entering a partially contaminated area from the contaminated area or entering into the buffer area from the negative pressure ward. It is absolutely forbidden to walk around casually with shoe covers.

2.6 The procedure for wearing protective articles when entering and leaving the isolation ward is as follows

Hand hygiene is required before putting on a mask when entering a contaminated area from the access area for medical personnel. Medical staff must only enter the isolation ward wearing PPE. About four times of hand hygiene are compulsory to be done when you leave the contaminated area to enter a clean zone. Currently, the requirement of so many times of hand hygiene might bother medical staff because the repetition is fussy. However, hand hygiene is of great importance to ensure the safety and an easy way to lessen the risk of infection. Therefore, certain times of hand hygiene are necessary, and it must perform qualified hand hygiene each time.

3. Nosocomial infection control

The phenomenon of nosocomial infection breaking out and a large number of medical care personnel becoming infected is related to the outbreak of community-acquired infection along with the failure in prevention and control of nosocomial infection. The confidence of governments and society in the control of the epidemic will be greatly undermined if patients get cross-infected or medical staff get infected in medical institutions.

The National Health Commission, on January 22, 2020, issued the Technical Guide for Prevention and Control of Novel Coronavirus Infected Pneumonia in Medical Institutions (First Edition) (hereafter this text will be abbreviated as Guide). This is divided into four parts, and the first part is the basic requirements.

There are 10 basic requirements for prevention and control of the novel coronavirus infected pneumonia in medical institutions.

(1) *Formulate an emergency plan and workflow.* Medical institutions at all levels are required to formulate an emergency plan and workflow for novel coronavirus infected pneumonia.

(2) *Train all staff.* In addition to medical staff, management, logistical, property, and cleaning personnel should also receive training to be aware of the emergency plan and workflow, and know how to prevent and control the epidemic. The personnel in nonhigh-risk departments should particularly take this training seriously and use standard precautions.

(3) *Enhance occupational protection of medical staff.* Some medical staff have become infected while this plan was being made, which is related to treatment operations or inadequate protection in hospitals. Therefore, medical institutions should standardize the work of disinfection, quarantine, and protection, and stock up on protective equipment. The Guide particularly emphasizes strengthening the control of contact, droplet, and air transmission based on the strict implementation of standard prevention. The correct selection and wearing of masks and hand hygiene are key measures to contain the epidemic. As a respiratory infectious disease, SARS-Cov-2 cannot cause infection until it enters the respiratory tract through droplets or aerosols. Nothing is more important than the mask—a firewall in front of the mouth and nose. As COVID-19 can spread through close contact, hand hygiene is particularly necessary. Hand hygiene is required when you want to touch any part of your body with your hands. The basic method of hand hygiene is to wash your hands with running water and hand sanitizer or to wipe your hands with instant hand sanitizer. Washing hands effectively cannot be finished in a few seconds and should continue for about 40 s; otherwise the hands will not be washed thoroughly. Quick hand disinfecting needs to last for 15–20 s, and can be achieved only with adequate instant sanitizer.

(4) *Pay attention to the health of medical personnel.* ① Deploy human resources and arrange shifts reasonably to avoid overwork of medical staff. Alternate work with rest and recreation, and ensure sufficient sleep. ② Provide a nutritious diet to enhance the immune systems of medical staff. ③ Carry out active health monitoring based on staff's characteristics of post and corresponding risk assessment results, including body temperature and respiratory symptoms. Pay attention to the prevention of colds and

infection, and improve fitness. Medical staff displaying symptoms such as fever and cough must receive immediate medical treatment, and not work while they are experiencing illness. ④ Take appropriate measures to ensure that medical staff provide medical services to patients safely. More than 3000 medical staff have been infected with COVID-19 in China, which mainly occurred in the early stages of the outbreak, for a number of reasons: lack of understanding of this emerging infectious disease; failure to identify infection early, inadequate knowledge of self-protection; improper use of protective equipment, busy work; and so on. Among these infections, some were caused by household clustering, some were from other community environments, and some were nosocomial.

The following matters need attention: ① Behavioral quarantine is tantamount to or even more important than PPE. Every detail counts. ② Put on and take off PPE in appropriate areas. Do not walk around in contaminated PPE. ③ Post the protocol of putting on and taking off PPE on the wall, and equip a full-length mirror. ④ Ensure hand hygiene by supplying sufficient instant hand sanitizer. Wash hands before leaving wards or eating meals. Do not touch your face with your hands. ⑤ Check the tightness of masks. Wear the appropriate mask in different areas. ⑥ Open windows to provide ventilation.

(5) *Strengthen the supervision of infection.* Monitor the infection of medical staff and confirmed suffers.

(6) *Carry out the management of cleaning and disinfection.* ① Strengthen the ventilation of diagnosing and treating environment according to the Management Standards of Air Purification in Hospital. Medical institutions should disinfect air or be equipped with air purifiers if possible. ② Implement the technical standards of disinfection in medical institutions strictly. Clean and disinfect the environment of diagnosis and treatment (including air, object surfaces, ground, etc.), medical devices, patients' items, patients' secretions and vomit, etc. Take scientific disinfection measures, and devise a precise strategy according to the relevant laws and regulations to cut off transmission routes and eventually contain the epidemic. Meanwhile, excessive disinfection should be prohibited, because it not only fails to increase the effect of disinfection, but also threatens public health.

(7) Strengthen the management of patients to avoid cross infection.

(8) Enhance the health education of patients, including the use of masks, hand hygiene, and the practice of social distancing.

(9) Strengthen the management of infection outbreak.

(10) *Strengthen the management of medical waste.* Contaminated masks, gloves, shoe covers, caps, gowns, and protective clothing should be put in the right places. Do not discard or scatter medical waste during its transportation.

In terms of key hospital departments, there are four main departments. The first department is the fever clinic where standard prevention, hand hygiene, timely detection (diagnosis) of cases, layout, division, and ventilation should be given attention. The second is the emergency department in which, besides layout, division, and ventilation, there are two points of importance: ① preliminary checking and differentiating diagnosis (direct fever patients to the fever clinic for medical consultations) and ② standard prevention. The third is the general ward, in which standard prevention is the most important. The medical staff must wear masks and practice hand hygiene correctly, otherwise they might become close contacts or infected. Moreover, infected patients must be identified in a timely fashion. The fourth department is the isolation section, in which suspected or confirmed patients with COVID-19 are received and treated. Here, layout, division, ventilation, and work schedule are of great importance. There should be three areas and two passages: a clean area, a partially contaminated area, and a contaminated area; and a corridor for patients and a passage for medical staff. Substantial barriers are required between the three areas. The air current direction should be from the clean area to the contaminated area. The negative pressure room should comply with relevant regulations, if any. Two further points must be noted: ① suspected cases must be isolated in a single room and ② confirmed cases can share a room.

The protocol and pier glass are required while wearing and taking off PPE. Medical staff should supervise and guide each other in order to improve together and protect themselves better.

3.1 Collecting and processing laboratory samples of patients with suspected COVID-19

Generally, there are two aspects. ① All laboratory samples should be considered to be potentially infectious, including blood, lavage fluid, nasopharyngeal swabs, and other body fluids. ② The medical personnel who collect, process, and transport any clinical samples should strictly abide by the following standard control measures and biosafety operations to minimize the possibility of exposure to the pathogens.

Standard control measures and biosafety operations:

(1) Ensure the medical personnel who collect samples use appropriate PPE (such as protective glasses, medical masks, protective clothing with long sleeves, gloves, etc.). If aerosol might be generated in the process of collection, such as the collection of alveolar lavage fluid and sputum aspiration, the collector should wear a special mask (NIOSH-certified N95 respirator or mask achieving EU-certified FFP2 standards, or mask at equivalent protective level).

(2) Ensure all personnel responsible for transporting samples are trained on the safety operation procedures and leakage disposal, which are hard to achieve in most medical institutions due to their wide coverage.

(3) The delivered sample should be placed in a leak-proof sample bag (i.e., a secondary container), which is a separate sealed sample bag (i.e., a plastic biohazard sample bag). During transportation, the secondary container should be placed in a sample container (i.e., a primary container) marked with the patient's information and attached with a clearly written laboratory examination application form.

(4) The laboratory in medical institutions must comply with the corresponding biosafety operation specifications and transportation requirements for different types of biological samples. For example, if the standard operation is not performed for samples possibly containing the virus, it is very likely to cause cross infection in the laboratory.

(5) All samples should be delivered manually whenever possible. A pneumatically driven pipeline system should not be used to deliver samples, because if the sample is overturned in the delivery pipeline, cross infection might happen in the hospital.

(6) The full name and date of birth of each patient and the warning "suspected novel coronavirus infection" must be clearly recorded on the laboratory examination application form.

Finally, once the sample is delivered, notify the laboratory as soon as possible to avoid sending samples to the wrong places and prevent problems during the process of delivery.

To strengthen the management of patients, the following aspects should be dealt with carefully:

(1) Suspected or confirmed patients should be isolated in good time and guided to the isolation area through the specified standard route by related staff.

(2) The patient should put on a hospital gown before entering the wards. Their items and clothes should be stored in a designated place and kept by the medical institution after being disinfected.
(3) Educate patients to choose and wear masks, and to follow coughing etiquette and ensure hand hygiene correctly.
(4) Strengthen the management of persons who visit or accompany patients. (For example, each hospitalized patient is allowed to be accompanied by up to one family member; no other person is allowed to visit suspected or confirmed patients.)
(5) For isolated patients, in principle, their activities should be limited to the isolation ward. If they need to leave the isolation ward or isolation area, they should wear medical surgical masks to prevent contaminating other patients or the environment.
(6) When suspected or confirmed patients are discharged or transferred to another hospital, patients should change into clean clothes before leaving and their wards should be disinfected.
(7) The bodies of deceased patients should be disposed of promptly. The method is: fill all open channels of the patients' body (mouth, nose, ears, anus, etc.) with 3000 mg/L chlorine disinfectant or 0.5% peracetic acid cotton ball or gauze, wrap the body with a double layer of cloth, and then put the body into a double-layer corpse bag and send the body directly to the designated place for cremation by specific vehicle. Personal belongings of patients during hospitalization could be cremated with the patient or taken home by their families after being disinfected.

4. Fangcang shelter hospitals/temporary treatment centers

Fangcang shelter hospitals are regarded as a key step at a critical moment. The establishment of Fangcang shelter hospitals is one of the important reasons why this epidemic has been controlled in China. Fangcang shelter hospitals contribute to the rapid examination and isolation of suspected cases and mildly affected patients, reducing the chance of human-to-human transmission in the community, and thus play a pivotal role in slowing down the increase of cases.

In the future, the construction of massive exhibition centers, gymnasiums, warehouses, and factories should leave interface and corresponding auxiliary space outside the interface, which are not used at ordinary times, but can be quickly transformed into a temporary treatment hospital in a short period of time when needed. The significance of Fangcang shelter hospitals is that in the national and even world emergency system, they can be constructed quickly to provide large numbers of medical beds and be operated at low cost. In terms of isolation, treatment, and supervision of patients with mild cases, measures such as Fangcang shelter hospitals also provide practical and significant experiences for other countries.

5. Disease control on a community level

Because there are neither corresponding vaccines nor specific drugs for COVID-19, currently the only effective measures to curb the epidemic are strict social prevention and control, such as the practice of social distancing, wearing masks, and washing hands frequently. If we had not taken strict measures, like the blockade of Wuhan, the situation in foreign countries would be worse right now, and there might be more than one "Wuhan" in China. The Chinese and English reports released by the Chinese expert group and the WHO expert group after the joint investigation clearly described China 's experience of prevention and control, with particular emphasis on not placing obstacles to nucleic acid testing—carrying out testing for any possible cases.

Faced with the severe COVID-19 epidemic, only by getting ahead of prevention and control at the grassroots level can we finally reduce the number of confirmed cases, critically ill patients, and deaths. In this outbreak, the general hospital receives and treats critically ill patients, while screening suspected cases, managing discharged patients and close contacts at home or centralized isolation, and health education of patients is mainly implemented at the grassroots level. The division of labor between the grassroots and general hospitals played a critical role in curbing this outbreak.

Grassroots medical personnel mainly undertake the following tasks: pre-examination and triage; timely detection, isolation, and referral of patients with fever and suspected cases; network management and carpet screening of the community; management of at-home or centralized quarantined patients; follow-up of discharged patients; strengthening education and guidance on the prevention and control of COVID-19; precise management of patients with chronic diseases (follow-up services, implementation of long prescriptions, substitute medicine-taken, home delivery services); and assisting other grassroots institutions in their work and giving professional guidance.

5.1 Use information and communication technology to get ahead of prevention and control

(1) Fully carry out registration, screening and reporting of community residents. Use direct reporting systems to support epidemic data filling and level-by-level statistics, focus on the coverage of suspected and confirmed cases, and continuously improve the efficiency and quality of data reporting. Use mobile apps, WeChat, and other forms to complete the registration of residents' travel information and health status. Set up a "high-risk reminder" in the information system to focus on and track high-risk and suspected people, and master the regional control situation comprehensively and dynamically. Use the Internet information system to screen, prediagnose, and triage patients with fever, and immediately refer to the nearest higher-level hospital with fever clinic if necessary. The information should be reported promptly after completing registration.

(2) Implement the information registration and reporting of home observers: ① Patient management during medical observation. Use network resources to assist in tracking and supervising home medical watch of personnel from epidemic areas. Within 14 days of medical observation, monitor their health status through network systems. For example, patients or their family members can report personal changes in time in the form of "electronic patient log card" through WeChat, mobile app, network registration, or telephone. Community health workers conduct real-time monitoring and report abnormal situations promptly. ② Management of the elderly and patients with underlying disease. Gradually realize the "remote bed monitoring system." Grassroots doctors use the internet to strengthen online precise management and services and focus on the "online health management" for the elderly and patients with chronic diseases such as hypertension and diabetes within their jurisdiction.

(3) Improve the efficiency of community health services with an intelligent voice system. Qualified communities can use the intelligent voice follow-up system to assist in various epidemic control work.

(4) Use "Internet +" to expand the scope of medical services: ① Online medical inquiry platform. Establish an online medical inquiry platform to provide consultation services for community residents. ② Online guide for medical attention. Construct an online platform which introduces the regional official registration platform: providing third-party services such as fever clinic inquiry and online registration to help residents conduct online self-assessment before visiting the fever clinic. If a resident has fever

symptoms, they could also enquire about the fever clinic nearest to their home and patient-guiding services through the platform in advance.

(5) Construct the correct guidance of public opinion, dispel rumors, and create an online science education platform.

(6) Provide online courses for basic level medical staff.

5.2 Improve the effectiveness of diagnosis and treatment with information technology

(1) Realize remote diagnosis and treatment. Take advantage of major hospitals and build a multilevel interactive service platform for diagnosis and treatment. Major hospitals provide services such as remote consultation and prevention guidance, and utilize expert resources with information technology to improve the capacity to deal with epidemic situations of grassroots medical institutions, ease the pressure of diagnosis and treatment in designated hospitals, and reduce the risk of cross-regional transmission. Construct remote diagnosis and treatment service for digital images and form a daily procedure of "remote consultation of suspected cases."

(2) Use a "medical consortium" to promote multilevel linkage interaction. Use the "Internet +" information platform to form a medical consortium of third-level hospitals in the same area with second-level hospitals, community hospitals, and township health centers in order to strengthen the exchange and sharing of information between the primary medical units and each superior medical unit. Make full use of the "novel coronavirus infected pneumonia sharing platform" constructed by the state and the data platforms in various regions to promote the effective linkage of multilevel medical institutions.

5.3 Construction of prevention and control platform

(1) Screen suspected patients, understand transmission for infection, and find "super spreaders." Collect and disseminate relevant information: the information of those who have been to the epidemic area, including the time of arrival in the epidemic area, the people contacted in the epidemic area, the way out of the epidemic area, and the places they have visited. Collect suspected symptoms: relevant residents upload their health status on the platform (whether they have fever, cough, expectoration, fatigue, diarrhea, and/or other symptoms). The abovementioned data can be conducive to screening suspected patients, understanding the transmission route, and

finding super spreaders, which is of great significance for prevention and control of the epidemic.

(2) Cut off transmission routes and circumvent the source of infection. Through an epidemic map, in which the real-time locations of suspected and confirmed patients are shown, the public can avoid the source of infection on their own initiative.

6. Vaccine development

What if COVID-19 becomes a "long-standing disease"? This is a matter of great concern at home and abroad. The development of vaccines is complicated; however, vaccine is an indispensable means of epidemic prevention in the battle against a persistent communicable disease.

CHAPTER 7

Better understanding and control of new coronavirus infection

Contents

1. How to better understand the 2019 coronavirus disease

We have been continuously deepening our understanding of 2019 coronavirus disease (COVID-19)—an emerging disease. Further knowledge on varying clinical manifestations, phenotypes, clinical course, acute and chronic conditions, susceptibility, as well as research to improve our ability in identification of susceptible populations and tracking the direction of evolution of the virus, are urgently needed.

COVID-19 has a wide spectrum of clinical manifestations, which are not limited to common respiratory tract symptoms, but also include various systemic reactions, such as fatigue and diarrhea. The reason why the World Health Organization named it 2019 coronavirus disease and not some kind of pneumonia or respiratory syndrome is that, in addition to the pulmonary involvement, this novel coronavirus also causes damage to multiple organs. Our accumulated experience on managing this emerging infectious disease is still far from sufficient.

COVID-19

https://doi.org/10.1016/B978-0-12-824003-8.00007-3

The clinical manifestations, clinical course, prognosis, severity, and treatment response of the disease vary from individual to individual. Moreover, the virus itself has not yet entered a stable state. After being transferred to a human host, the virus will strive to adapt to this new host by evolution and mutation. How the interaction between the virus and the human body determines disease severity and acute and chronic processes, including persistent infection, requires further investigation. For now, many puzzles still exist regarding the disease in clinical practice. For example, whether different populations, interventions, or blocking methods will affect the adaptation, mutation, and evolution of this novel virus, and why hosts with different genetic characteristics have different responses to it, remain unclear. Therefore, we need to pay attention to some detailed and in-depth information on the disease at the early stage.

The coronavirus itself has a lot of complexity. It is unknown how this kind of pathogen will evolve or mutate next. Considering that the new coronavirus has an 80% nucleotide sequence similarity with SARS coronavirus, the International Committee on Taxonomy of Viruses officially named it SARS-CoV-2. From a biological perspective, although two pathogens have a homologous relationship, their phenotypes and biological natures are quite different when there is a 20% difference in nucleotide sequence. A huge knowledge gap remains on the immunogenicity of SARS-CoV-2 and its relationship with its host. SARS-CoV-2 is transferred from a natural host to the human body through a certain route to cause infection. Afterwards, the virus is prone to mutate or evolve for adaptation to the new host, which is the so-called "host adaptation" in biology.

Perhaps the virus' ideal state is showing persistence or enhancement of transmissibility, but with relatively attenuated pathogenicity. If so, it can exist in the host for a relatively long period of time. The change of the pathogen itself is not only a dependent variable that modifies as the environment changes, but also an independent variable that affects disease conditions of the host. The interactions between the pathogen and the host contribute to different manifestation of the novel disease, which calls for special attention and improved understanding. It remains unclear in which direction the pathogen will evolve. Viruses of different origins have varied survival environments and survival opportunities. For example, all kinds of treatments, intervention measures, and hosts will make a virus change differently.

To sum up, we must fully recognize that after the new pathogen enters the new host of mankind, both pathogen and the host will change in an interactive manner, which is an important feature of this emerging disease.

The understanding of the occurrence and development of this novel disease in terms of viremia, ARDS, SOP-like change, hypercoagulability/fibrinolysis and VTE, myocardial injury and its related markers, AKI, and so on has been mentioned before.

In short, as COVID-19 is a newly occurring disease, clinical and scientific data are still limited, driving further exploration. It is necessary to master some basic knowledge on diagnosis and scientific laws of this disease. For example, nucleic acid detection should be carried out in multiple specimens from different body sites or body fluids simultaneously, including pharynx, saliva, lower respiratory tract secretions, blood, and anus; for antibody detection, IgM serves as an acute phase antibody, and IgG is used as a convalescent antibody.

2. Improving clinical management of respiratory tract virus infection and future research agenda

First, it is necessary to recognize the high frequency and threats of respiratory tract viruses; second, we need to establish a clinical-oriented detection platform; third, we must pay attention to the rational selection and application of antiviral drugs; fourth, we should conduct clinical and basic researches on respiratory tract viruses.

2.1 Detection of respiratory tract viruses: Etiological diagnosis of lung infection and emerging infectious diseases

First of all, identification of etiology could help guide medical treatment decisions. Second, ensuring early diagnosis and treatment of viral infection improves prognosis compared to initiating treatment later. The etiological examination is indispensable to achieve early treatment. Respiratory virus tests are necessary for reducing inappropriate use of antibiotics, surveillance of emerging infectious disease, and exploration of new prophylactic measures.

(1) Influenza detection, diagnosis, and treatment process. The whole process will be performed according to the demands of diagnosis and treatment, so it is necessary to determine whether patients manifest symptoms of influenza. For hospitalized patients presenting influenza manifestations, it is recommended to start empiric antiviral treatment and detection of influenza virus simultaneously. Judging whether hospitalization requirements and high risk factors exist helps us to conduct virus testing and prescribe medication.

(2) Experience of respiratory tract virus detection. Currently, the existing rapid antigen detection reagents are not sensitive enough—PCR is still the gold standard diagnostic method. The positive detection rate in throat swabs is lower than that of lower respiratory tract specimens, especially in patients infected with avian influenza virus or SARS-CoV-2. Convalescent antibody is used for further diagnosis. IgM antibodies appear 3–7 days after the onset of COVID-19. Considering that virus isolation has a low positive rate, it only serves as an important research tool and should not be used for clinical diagnosis. Virus detection should not be limited to influenza viruses or novel coronavirus due to the possibility of infections caused by other respiratory tract viruses.

2.2 Establish a respiratory tract virus research platform for clinical diagnosis and treatment

Confirmed diagnosis requires clinical-oriented detection methods. Currently, many new molecular biological diagnostic methods are available, owing to the huge support of government to help many research teams carry out multidimensional basic research and explore new detection methods during the epidemic outbreak. For example, multiple PCR microfluidic technology (Film-Array and GeneXpert) could obtain detection results within 1–1.5 h. There are also other new diagnostic techniques, such as chip technology (Verigene System and shinychip), matrix-assisted laser ionization resolution time-of-flight mass spectrometry (MALDI-TOF), Fourier transform infrared spectroscopy (FT-IR), second- and third-generation sequencing technology, as well as point-of-care testing. These techniques are expected to achieve "four high" requirements: high throughput, high speed, high sensitivity, and high precision. However, many of them are still in development or the early phase of clinical trials. Although the development of new detection techniques is not the area of expertise of our clinicians, the specificity, sensitivity, and influential factors of these methods in clinical application are the issues that clinicians should take into consideration and have a good command of. Relevant clinical research should be carried out, so that we can correctly interpret the results and guide clinical diagnosis.

We should ensure that the specimens are correctly collected and transported according to the appropriate standards, the laboratory test is conducted promptly, and the result is consistent with actual clinical conditions. The combination of respiratory tract virus detection results with clinical diagnosis and treatment is essential to improve both detection and clinical diagnosis levels.

It has been gradually realized that multiple sampling at different time periods during the whole clinical course and conforming to a standardized process can improve the positive rate of virus detection. In order to reduce repeat tests, we performed a single testing by mixing throat swabs and nasal swabs together rather than separating the specimens for individual detection tests. It turned out that this method yields a higher positive rate and is more cost-effective. Therefore, detection of multiple specimens at the same time can improve the positive rate.

For influenza, prompt intervention and medication can reduce the risk of developing into severe cases and deaths. Therefore, early diagnosis cannot be emphasized enough due to its critical influence on patients' prognosis.

2.3 Attach importance to rational selection and application of antiviral drugs

When using antiviral drugs, etiology, medication (drug classes, dosage, duration), and the application in special populations should be taken into account. Ten years ago, only a few antiviral drugs were available for clinical treatment. Currently, some newly developed antiviral drugs have been put into use, but there are still limited classes of drugs. It is estimated that there will be emerging antiviral drugs in the next decade.

2.4 Establish a comprehensive treatment platform for severe cases' life support

ARDS, sometimes accompanied by multiple organ failure, is the most common reason for the lethality of viral pneumonia. Therefore, it is essential to establish a comprehensive treatment platform providing life support to severe cases, which also helps to control the epidemic.

Respiratory support should be graded according to the severity of the disease: at the early stage of disease, oxygen support or high flow oxygen therapy is recommended; when the disease progresses and dyspnea occurs, switching to noninvasive ventilation is advised; and severe patients may receive tracheal intubation for invasive ventilation.

During the outbreak of SARS, although there were no prospective controlled studies showing what kind of ventilation pattern and mask would be more effective, nasal CPAP masks were extensively used for treatment due to the following considerations. First, the noninvasive ventilation in SARS patients is continuously applied for a few days or even 1–2 weeks, among which the nasal mask achieves the best patient compliance. Second, CPAP does not require man-machine synchronization, and the nasal CPAP mask is therefore preferred for 24-h continuous use, although there is no recommendation from evidence-based medicine.

Fig. 1 shows an H1N1 influenza patient admitted to the respiratory and critical care medicine department of the Guangzhou Institute of Respiratory Health. The patient was supported by ECMO because ventilation alone no longer helped. Managing such a critically ill patient is very challenging. The patient received ECMO support for 1 month and invasive ventilation for 2 months, and finally was successfully extubated and discharged from hospital.

Critical patients sharing similar clinical presentations with the patient illustrated in Fig. 1 are supposed to have survival opportunity in hospitals with relatively good resources, and the change of those patients' long-term rehabilitation is usually good. There are many fibrous foci in the lungs on discharge, but patients can recover very well in 6 months to 1 year, and can even exercise and work normally. So for this kind of case, we should give comprehensive life support measures to reduce the mortality rate and ensure better long-term effects.

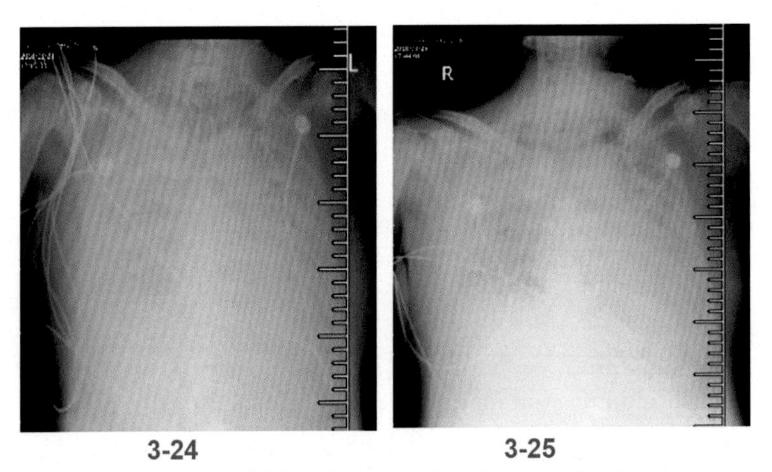

Fig. 7.1 Chest image of an H1N1 influenza patient (provided by the respiratory and critical care medicine division of Guangzhou Institute of Respiratory Health).

2.5 Develop clinical and basic research in respiratory tract virology

Carrying out scientific exploration and research on the respiratory tract virus with a combination of basic research and clinical work will provide guidance to improve our performance in clinical diagnosis and treatment. Clinical issues including early diagnosis, prompt intervention, early warning of developing into severe cases, and treatment of critically ill patients are all worth investigating. For example, clinical trials should be conducted to determine the specificity, sensitivity, positive predictive value, and negative predictive value of some newly developed detection methods, to establish a standard screening process, to optimize and explore the effects of early application of antiviral drugs, and to explore predictors of developing into severe cases (e.g., viral load, the dynamic change of inflammatory mediators). Patients at risk of progressing to severe cases are very likely to develop into critical illness that requires mechanical ventilation if they do not receive early treatment. Exploring specific treatments, especially convalescent plasma therapy, is urgently needed to prevent such patients from developing into critical cases. Managing severe cases requires comprehensive life support measures, and rigorous evidence-based clinical researches are awaited to guide the reasonable and appropriate use of respiratory support, including the optimal time period to initiate noninvasive ventilation, endotracheal intubation, and ECMO.

3. The need to conduct clinical research in an orderly way

One of the important hallmarks of medical maturity in China is the capability to perform diagnosis and treatment with rigorous standards based on the accumulated experience from past medical history, and to conduct research with great concentration. Considering the large number of current clinical researches, it is vital to strengthen coordination, and guarantee the quality of the implementation. A large amount of literature review and a preliminary observation of 2019 coronavirus disease are necessary for a good research design.

It is expected that in the future, clinical research can be carried out in a more orderly way. It is hoped that Chinese medical practitioners can stand at the forefront of international medical science in confronting clinical emergencies.

4. The importance of PCCM discipline development to cope with the epidemic situation

There is a general consensus among the firstline respiratory and critical care medicine specialists all over China that the tightly combined development of respiratory disease and critical care medicine, which is called pulmonary and critical care medicine (PCCM), is very important for coping with such an epidemic.

If there is no establishment and development of the discipline of PCCM, pulmonologists' understanding of critical care medicine will be insufficient, and their abilities in management of critically ill patients, treatment of respiratory failure and application of respiratory support will not match current performance. To treat COVID-19 patients with comprehensive therapies represented by antiviral drugs and respiratory support, it is necessary to combine etiological treatment and symptomatic treatment, and to have a deep understanding of the pathological changes of the respiratory system, respiratory physiology, and respiratory tract pathogens. The first group of colleagues who have completed PCCM training have acted with clear-minded determination and professionalism, becoming the most important backbone force among frontline fighters in the battle against the COVID-19 epidemic.

As the lungs are the major targets for SARS-CoV-2, most leading frontline experts are pulmonologists. After years of construction and development of PCCM, pulmonologists have shown professional capabilities in this public health emergency and the discipline of respiratory disease is transforming rapidly from scale enlargement to quality improvement. The COVID-19 epidemic is also a test for the achievements of our discipline construction and fellowship training program.

It is recommended that colleagues on the firstline response to COVID-19 from departments of respiratory and critical care medicine all over the country should have greater exchange of academic ideas and understanding of scientific laws of this disease. At the same time, we should focus on the future and take responsibility for preventing diseases from the perspective of discipline construction and fellowship training.

It remains unknown how long COVID-19 will last and whether it will return in the coming years. It has been the third time that coronavirus has endangered human health since the outbreak of SARS 17 years ago, and there is a significant possibility that other coronaviruses will emerge in the next few decades. Identifying coronavirus and determining its specific effect on immune function are particularly worthy of our attention. What we have learned from the past and should plan for the future include the development of clinical discipline, the construction of emergency and public health systems, and the improvement of the overall public health system, especially the integration of clinical and preventive medicine.

Index

Note: Page numbers followed by *f* indicate figures.

D

E

F

G

H

I

L

M

N

O

P

Printed in the United States
By Bookmasters